FROM STEAM TO DIESEL

PRINCETON STUDIES IN BUSINESS AND TECHNOLOGY

SERIES EDITOR

DAVID HOUNSHELL

Before the Computer: Mechanical Information Processing in the United States, 1865–1956, James W. Cortada

Politics and Industrialization: Early Railroads in the United States and Prussia, Colleen A. Dunlavy

FROM STEAM TO DIESEL

MANAGERIAL CUSTOMS AND
ORGANIZATIONAL CAPABILITIES
IN THE TWENTIETH-CENTURY
AMERICAN LOCOMOTIVE INDUSTRY

Albert J. Churella

PRINCETON UNIVERSITY PRESS   PRINCETON, NEW JERSEY

Copyright © 1998 by Princeton University Press
Published by Princeton University Press, 41 William Street,
Princeton, New Jersey 08540
In the United Kingdom: Princeton University Press,
Chichester, West Sussex
All Rights Reserved

Library of Congress Cataloging-in-Publication Data
Churella, Albert J., 1964–
From steam to diesel : managerial customs and organizational capabilities in the
twentieth-century American locomotive industry / Albert J. Churella.
p.   cm. — (Princeton studies in business and technology)
Revision of the author's thesis (Ph.D.)—Ohio State University, 1994.
Includes bibliographical references (p.) and index.
ISBN 0-691-02776-5 (cloth : alk. paper) 1. Locomotive industry—United States—
Management—History—20th century. I. Title. II. Series.
HD9712.U52C48 1998
338.4'762526'0973—dc21   97-51800
CIP

This book has been composed in Caledonia

Princeton University Press books are printed
on acid-free paper and meet the guidelines
for permanence and durability of the Committee
on Production Guidelines for Book Longevity
of the Council on Library Resources

http://pup.princeton.edu

Printed in the United States of America

10 9 8 7 6 5 4 3 2 1

Contents

Acknowledgments	vii
Introduction	3
I. Steam vs. Diesel: The Capabilities and Requirements of a Radically New Technology	10
II. Internal-Combustion Railcars: Springboard to Participation in the Diesel Locomotive Industry	23
III. First-Mover Advantages and the Decentralized Corporation	37
IV. ALCo and Baldwin: Established Companies, New Technologies	58
V. Policy and Production during World War II	75
VI. Postwar Dieselization and Industry Shakeout	95
VII. The Era of Oligopoly	127
Conclusion	146
Notes	155
Bibliography	201
Index	213

Acknowledgments

AS THIS BOOK is a revision of my doctoral dissertation, I am deeply indebted to my adviser, Mansel Blackford. Without his insightful guidance and patient understanding, my interest in this project would never have come to fruition. K. Austin Kerr and William Childs have also given generously of their time and advice in seminars and through individual consultation. I appreciate greatly the comments and suggestions made by David Hounshell and Steven Usselman. Their insights have helped me to see the shortcomings in my original dissertation as I have worked to transform it into a book. I have had the opportunity to develop my study of the locomotive industry in a recent article in the *Business History Review*, and I am grateful to that journal for permitting me to reprint portions of that article in this work. Conferences of the Business History Society, the Society for the History of Technology, the Economic and Business Historical Society, and the Ohio Academy of History have also provided a forum for my work during its formative stages.

This book has absorbed many hours of archival research. Unfortunately, since General Motors and General Electric are still actively competing in the diesel locomotive industry, their records are not open to scholars. Furthermore, when the old steam locomotive builders left the industry, railroad enthusiasts and historians often saved a treasure trove of locomotive photographs, drawings, and blueprints, but could not preserve correspondence, memoranda, reports, or other written sources. To the extent that they were able to do so, I am eternally grateful. In particular, I would like to thank railroad historians and experts Jim Mischke and Wallace Abbey for giving me expert technical advice on the locomotive industry and its products.

The research for this book was made possible by the generous assistance of librarians, archivists, and scholars throughout the United States. Those who have helped include Tab Lewis and James Cassedy at the National Archives in Washington; Gregory J. Plunges at the National Archives, Northeast Region; Beverly Watkins at the National Archives, Great Lakes Region; Elaine Fidler and Ron Wiercioch at the U.S. Department of Justice; Joyce Koeneman and Andrea Cheney at the Association of American Railroads Library; Mark Cedeck, Curator of the John W. Barriger III Collection at the St. Louis Mercantile Association Library; Charles Bates and John Keller at the Allen County (Ohio) Historical Society; Jim Bachorz at the American Locomotive Historical Society; Sharon Nelson at the Pennsylvania State Archives; George Deeming at the Railroad Museum of Pennsylvania; Connie Menninger at the Kansas State Historical Society; Denise

Conklin at the Pattee Library of the University of Pennsylvania; Christopher Baer, Lynn Catanese, and Marjorie McNinch at the Hagley Museum and Library; the staff of the George Arents Research Library at Syracuse University; the staff of the Minnesota Historical Society in St. Paul; Richard Scharchburg and Bruce Watson at the General Motors Institute; George Wise at General Electric; L. Dale Patterson at the University of Louisville; Bridget Burke at the Smithsonian Institution; Blaine Lamb at the California State Railroad Museum; Kay Bost, Curator of the DeGolyer Library at Southern Methodist University; Reese Jenkins, editor of the Thomas A. Edison papers at Rutgers University; Roland Marchand at The University of California at Davis; Leonard Reich at Colby College; and Richard Tedlow at Harvard University. All of these individuals have been more than generous in their support of my research and writing.

Research can often be an expensive proposition, especially when extensive travel is involved. Several grants have helped to reduce this financial burden. The Ohio State University Graduate School provided a Graduate Student Alumni Research Award, while the Ohio State University History Department awarded a Ruth Higgins Summer Research Fellowship. Finally, The University of Illinois Foundation generously provided a John E. Rovensky Fellowship in Business and Economic History.

This book would doubtless never have been written without the ongoing support and sympathy of my parents, Albert A. and Helen Churella. My greatest debt—one that I can not adequately express in words—must go to my wife, Marianne Holdzkom. In spite of research trips, writer's block, and numerous moves, she has exhibited compassion, understanding, and love beyond measure. Finally, while many individuals have contributed generously to my work, any errors or omissions are solely my own responsibility.

FROM STEAM TO DIESEL

Introduction

FOR MORE THAN a hundred years, from the 1830s through the 1940s, steam locomotives formed the main power source for railroads throughout the world. Beyond their duties as haulers of freight and passenger traffic, steam locomotives symbolized both the romance of the rails and the industrial might of the American economy.[1] For all of their undeniable power and majesty, however, steam locomotives were notoriously inefficient and costly to maintain. For all of their symbolic reference to American industry, steam locomotives remained customized, hand-built products in a nation that enthusiastically embraced mass production. Any form of motive power that could overcome the technological limitations of the steam locomotive would, as a radical technological discontinuity, have profound effects on established steam locomotive producers.

Diesel locomotive technology represented just such a discontinuity. Diesel locomotives not only made more efficient use of fuel than did steam locomotives, they also cost less to service and repair. Beyond that, diesel locomotive technology was much more amenable to mass-production methods and the savings that resulted from these manufacturing techniques. The technological advantages of diesel locomotives had become evident by the mid-1930s, and diesels quickly replaced steam locomotives in most railroad assignments. By the late 1950s, it was difficult to find a steam locomotive in revenue service anywhere in the United States.

This book does not explore the development of diesel locomotive technology, per se. Rather, it explores the ways in which corporations and corporate executives responded—or failed to respond—to technological change. The locomotive industry provides the backdrop for an exploration of the processes of innovation and technological diffusion, the role of managerial culture and tradition in decision-making processes, and the ways in which both random events and exogenous forces helped to shape competitive patterns.

The dieselization revolution that swept through the American railroad industry in the decade following World War II changed forever the structure of the locomotive industry in the United States. Three companies—the American Locomotive Company (ALCo), the Baldwin Locomotive Works, and the Lima Locomotive Works—had long controlled steam locomotive production. Although all three firms built diesel locomotives, they had abandoned locomotive production by the late 1960s. In their place, the Electro-Motive Company (later the Electro-Motive Division of General Motors) and General Electric came to dominate the American diesel locomotive industry.[2]

The transition from steam to diesel locomotive production thus offers an early example of a complete turnover of firms within a single major industry. As a part of this turnover, the size and profitability of both ALCo and Baldwin plummeted. In 1917, the two companies were ranked 52 and 62, respectively, among the two hundred largest industrial corporations in the United States. By 1948, their rankings had fallen to 145 and 143, and continued to decline in the years that followed. With only two exceptions (Great Western Sugar and Willys-Overland), no industrial corporations in the history of American business fell so far, so fast.[3] At the same time, two large diversified corporations came to dominate the diesel locomotive industry.[4]

Context and Scope of the Dieselization Story

The locomotive industry during the twentieth century provides a fascinating case study of technological change and industrial transformation precisely because this total firm turnover existed. The speed and completeness with which the Schumpeterian "gales of creative destruction" swept through the locomotive industry raises several important issues.[5]

The first set of questions concerns the source of technological change, diffusion, and progress. Technological discontinuities may arise from a variety of sources, such as the organization, a technological system, or a community of technological practitioners.[6] Historians and economists have often been reluctant to explore the root causes of the technological changes that in turn contribute to alterations in firm behavior.[7] The evolution of the locomotive industry provides insights into this often unexplored black box of technology.[8] Furthermore, activities inside the black box frequently occur in a random, chancy, and often irrational manner.[9]

Inasmuch as the diesel engine succeeded in railroad service only half a century after its initial development, after first making a wide detour through the realm of maritime motive power technology, the issue of technology transfer is of considerable importance to this case study. While lone individuals—the "heroic inventors"—played an important role in the development and diffusion of diesel locomotive technology, "communities of technical practitioners" were also able to impose their values on corporate hierarchies, at least during the early stages of the industry.[10] In particular, the locomotive industry indicates that the role of the "heroic" inventor (or, at least, the zealous promoter) may have been underemphasized in technological histories of large corporate entities.[11]

The history of the locomotive industry is also the story of innovation in both technology and business practices. Since other historians have ex-

INTRODUCTION

plored the invention of the diesel engine and since the subject is beyond the scope of this volume, innovation assumes greater weight in this study.[12] While success in the locomotive industry at first depended largely on product innovation, process innovation—the development of new operating routines, particularly in the realm of production controls and marketing initiatives—ultimately held more importance.[13]

Diesel locomotive technology, by its very nature, determined production methods and competitive patterns, suggesting that "soft" determinism and related historical paradigms such as technological momentum and path dependency had considerable importance in the locomotive industry. Conversely, stochastic elements, such as historical chance and luck, played a considerable role in the development of the locomotive industry and in the diffusion of diesel locomotive technology. Exogenous—and largely coincidental—technological linkages also affected the pace of technological change in the locomotive industry.[14]

This study also addresses the role of patents and proprietary information as either an inducement or a barrier to technological diffusion. Patents, in the end, were of scant importance in the locomotive industry, and ALCo, Baldwin, and Lima suffered few setbacks because they lacked access to proprietary information. At the same time, companies were often unable to guard against "leakages" from their stock of proprietary knowledge.[15] The locomotive industry thus differed from the American railroad industry, in which patent pools encouraged technological diffusion while doing little to protect the proprietary rights of the inventor.[16]

This work also explores the relative effectiveness of varied corporate responses to technological change. American locomotive builders employed a number of competitive strategies in response to the change in technology from steam to diesel motive power. These included joint ventures, such as that between ALCo and General Electric, horizontal combination, as in a merger between Lima and Baldwin, and vertical integration, in the case of General Motors. These varied organizational responses to technological change, in turn, relate to the larger issue of how established companies adapt, or fail to adapt, to technological, political, social, and economic changes.

The locomotive industry also offers the possibility of assessing the potential for large diversified firms to exploit technological innovation and thereby fulfill larger social and economic needs.[17] The financial and technical resources of GM and GE facilitated their capture of the diesel locomotive market, indicating that modern diversified firms may be particularly efficient at exploiting technological innovations. Companies often have naturally differing abilities to respond to technological change, and such asymmetries—whether based on firm size or on other factors—can cause profound changes in competitive patterns and market structures.[18] While the expansion of firms into related product lines may initially increase competition, this process may lead ultimately to a reduction in competition; and this, in turn, may produce economically or socially undesirable consequences. In particular, this work examines GM's effects on the structure of the locomotive industry in light of the Justice Department's intense scrutiny of that corporation during the 1950s and 1960s. A related issue concerns the degree to which firms—large or small—can exert control over their technology and their markets.[19]

While the locomotive industry never assumed such a high profile as, for example, automobiles or aerospace, the importance of railroads to the American economy, particularly in wartime, and the emergence of a large and potentially monopolistic firm in the locomotive industry ensured that government policy would intersect with technological change and business strategy. As such, the locomotive industry offers insights into the effects of government policy, both proactively and as an attempt to alter established competitive patterns.[20]

In recent years, an ongoing dialogue has developed between historians who examine mass production, such as Alfred D. Chandler Jr., and those who have turned their attention to smaller, more specialized producers of small-batch or custom-manufactured products.[21] This work will describe how the locomotive industry, between 1930 and 1960, evolved from flexible small-batch production methods to manufacturing techniques that more closely resembled mass production. In other words, ALCo, Baldwin, and Lima had to respond to a radical technological discontinuity by making equally radical changes in their style of production. Managers at all three companies proved understandably reluctant to abandon time-tested methods of small-batch production, yet, at the same time, they chose to remain in the locomotive industry, even though this industry increasingly devalued small-batch manufacturing techniques.

The ultimate inability of the three established steam locomotive producers to make the transition from small-batch custom production to standardized near-mass production in turn raises a larger issue; namely, the extraordinary difficulty of transforming firm routines to accommodate radical technological discontinuities. In the case of the locomotive industry, ALCo, Lima, and Baldwin mastered the art of making incremental technological changes to their established product lines. These changes tended to be competence enhancing—in other words, they encouraged the "big three" producers to reinforce their established organizational strengths (their core capabilities) and create higher barriers to entry, while many less successful competitors fell by the wayside, particularly during the nineteenth century. And, by the 1930s, ALCo, Baldwin, and Lima had so effectively eliminated Hughesian "reverse salients" in steam locomotive technol-

INTRODUCTION   7

ogy that steam locomotives were approaching the limits of their technological potential.[22]

Diesels certainly represented a radical technological discontinuity, since they did not share any significant technology or components with steam locomotives and since their manufacture demanded vastly different organizational routines and competencies. Furthermore, since diesels offered a threefold increase in thermal efficiency over steam locomotives, there was little incentive for railroads to retain the old technology. Radical discontinuities tend to destroy competency, and this was certainly the case in the locomotive industry. The technological capabilities and organizational routines that had worked so well for so long in the steam locomotive industry were ill-suited to the vastly different requirements of steam locomotive production. Efforts to transfer basic design concepts (i.e., reliance on horsepower, rather than tractive effort; attempts to design a single high-power unit rather than linked low-power locomotives; and reliance on casting rather than welding) from one industry to the other created serious difficulties for steam locomotive producers in the diesel locomotive market.[23]

At the same time, new entrants (particularly Electro-Motive) proved especially adept at managing this radical technological discontinuity and creating substantial first-mover advantages. Electro-Motive enjoyed the classic "attacker's advantage" because that company had no financial, physical, or human resources invested in the core organizational capabilities of steam locomotive production.[24] Furthermore, Electro-Motive was free to explore a new set of performance characteristics and market applications (what Clayton M. Christensen and Richard S. Rosenbloom define as a "value network") for diesel locomotive technology.[25] In other words, even though early diesels could not compete against steam locomotives under the set of performance criteria understood by the steam locomotive industry (high horsepower, low cost), they offered characteristics (operating efficiency, flexibility, relative cleanliness) that steam locomotives could not match. Even though two of the established steam locomotive builders had developed diesel locomotive technology before Electro-Motive entered the field, these efforts bore little fruit, in part because managers at ALCo and Baldwin saw quite clearly that these primitive machines were woefully inadequate, based on their traditional standards of evaluation.

The difficulty and sluggishness with which managers at established firms responded to the emergence of a new value network is often grounded in corporate managerial culture.[26] In any industry, if finances permit, physical facilities can be reconfigured, workers can be retrained, and new technologies can be developed or purchased. This process, difficult and expensive under the best of conditions, becomes much more complicated if a company is infused with a corporate managerial culture that places undue reliance on traditional product lines and production processes.[27] Although executives are rarely locked into a rigid corporate culture, and their companies are seldom locked into rigid technological trajectories, both corporate culture and technological trajectories constrain the ability of individuals and firms to respond to technological change.[28] Likewise, since corporate culture reduces flexibility on an individual level, technological paradigms often have the same effect on a corporate level.[29]

Executives at ALCo and Baldwin, through lifelong training and experience, developed a corporate culture that was virtually inseparable from the custom craft production of steam locomotives. This corporate culture allowed ALCo and Baldwin to manage incremental changes in steam locomotive technology and thus dominate the steam locomotive industry and gain respect throughout corporate America as successful, reliable, and well-managed firms. Paradoxically, the corporate culture that contributed to this success had become ossified by the 1920s, with managers slow to respond to growing evidence that the steam locomotive industry was headed for extinction.[30] The very success of their corporate culture blinded executives at ALCo and Baldwin to the opportunities—and requirements—of radical technological change.[31]

At the same time, executives at Electro-Motive possessed a corporate culture that, while not inherently "better" than that of ALCo and Baldwin, nonetheless recognized the applicability of the diesel locomotive and fostered its development. In particular, executives at Electro-Motive, unlike their counterparts at ALCo, Baldwin, and Lima, realized that advances in diesel locomotive technology greatly increased the importance of advertising, marketing, and post-sales support services. The success of a company's marketing efforts thus had as much effect on its profitability and long-term survival as did the applicability of its technology. And, while corporate culture was embodied in firm routines at the steam locomotive producers, it was largely independent of them at Electro-Motive—thus furthering the new entrant's "attacker's advantage."[32]

The locomotive industry offers ample evidence of corporate success and corporate failure. While a variety of factors, ranging from issues of technological transfer to instances of government action, influenced changing competitive patterns in the locomotive industry, endogenous factors proved to be the determinant of success or failure in the industry. As such, the industry offers examples of both the successes of individual initiative and of the limits of bureaucratic managerial control over the use of technology. In the locomotive industry, at least, deeply rooted corporate cultures ensured that the development of technology within the black box and corporate decisions regarding the applicability of that technology did not proceed predictably, or in a manner that could be consistently and effectively subsumed under a corporate hierarchical decision-making process.

INTRODUCTION

The response of the American locomotive industry to technological change helps to illuminate the diverse responses of American business to the forces of change. The decline of established, respected, and profitable locomotive producers and the ascendancy and subsequent financial success of two new entrants into the field offers insights into the decline of established American industries and the rise of new industrial powers and new forms of economic endeavor.

I

Steam vs. Diesel: The Capabilities and Requirements of a Radically New Technology

STEAM AND DIESEL locomotives embodied vastly different technologies and, of even more concern to locomotive builders, these technological differences mandated radically different production and marketing techniques. In particular, even slight variations in operational requirements could require steam locomotive designs to be altered substantially. As a result, steam locomotive designs proliferated, and builders constructed steam locomotives in small batches, customized to suit the requirements of a particular railroad or operating district. Steam locomotives remained customized, purpose-built machines, and the necessity of tailoring locomotive designs to specific railroad requirements ensured that economies of scale were largely unobtainable and that customers would have the upper hand in the design process.

Diesels were more adaptable, however, ensuring that fewer designs could serve a wider variety of applications. This versatility, combined with high research and development costs, meant that diesels could be most efficiently produced in large batches to standardized designs. In other words, the shift from steam locomotive production to diesel locomotive manufacturing not only involved a radical transformation in technology; but that technological transformation also required companies and their managers to shift from customized, small-batch production techniques to the philosophy and methods of mass production.

Traditional Steam Locomotive Industry Practice

By the 1920s new steam locomotives were far larger than those built fifty years earlier, yet production techniques had changed little during the interval. Numerous builders produced small industrial steam locomotives; and some railroads built locomotives in their own shops. By the early years of the twentieth century, however, only three firms produced large mainline steam locomotives for independent railroad customers on a regular basis. Baldwin traditionally held a leadership role in the industry, although it traded market dominance with ALCo at regular intervals. Lima, a marginal producer, often accommodated the overflow from its two larger rivals. Between 1920 and

1928, ALCo enjoyed a 47 percent market share, followed by Baldwin, with 39 percent, and Lima, with 14 percent.[1]

While the capital equipment sector traditionally exhibited close working relationships between suppliers and customers, this process was rarely more apparent than in the steam locomotive industry.[2] Railroad executives, particularly those in motive power departments, demanded custom locomotive designs, and none more so than motive power experts. Especially during the nineteenth century, this quest for uniqueness partly resulted from the vanity of motive power officials who often believed that their designs were superior to those employed by their competitors. Railroads possessed differing operational characteristics (flat vs. mountainous terrain, freight vs. passenger traffic, impeccably maintained mainline vs. lightly built branchline), and this provided a more rational basis for custom orders. Despite design multiplicity, the steam locomotive builders, notably Baldwin, succeeded admirably in their efforts to standardize production techniques, primarily by using as many common components as possible for each locomotive design. The standardization of components did not extend to the locomotives themselves, however, and ALCo, Baldwin, and Lima produced a bewildering array of locomotive models. Baldwin, for example, offered 492 designs in 1915.[3]

ALCo, Baldwin, and Lima produced major locomotive components in separate buildings or areas, such as the tender shop, the frame shop, and the boiler shop. In these buildings, skilled craftsmen created parts and subassemblies to precise specifications, yet manufacturing tolerances were typically somewhat "loose" and parts often varied by as much as 1/100th of an inch. Skill and success in forging, casting, and machining ultimately determined the quality of locomotive components. At the end of the production process, locomotive components came together in the massive erecting shop, where detailed "erecting cards" specified how some six or seven thousand separate parts would fit together.[4] In the erecting shop, skilled craftsmen built, rather than assembled, the locomotive piece by piece. In 1939, an observer wrote that one builder, Lima, had a "pride in craftsmanship that has never been sacrificed to a conveyor belt . . . ," at a time when the conveyor belt had already transformed the automobile industry, and many other industries as well.[5]

As was typical of many companies in the producer-oriented capital goods industry, the three steam locomotive producers possessed limited marketing capabilities. Top-level managers worked diligently to secure steam locomotive orders, relying on their close personal contacts with railroad motive-power officials. Builders generally solicited competitive bids from the railroads and were required to do so after passage of the Clayton Act in 1914.[6]

Steam locomotive builders offered little in the way of financing or postsale support services. Baldwin, in particular, refused to provide financing for its steam locomotives, informing one customer that "as you can appreciate, the Baldwin Locomotive Works as an equipment manufacturer is not in a position to undertake such financing for its own account."[7] Instead, railroads typically sold equipment trust certificates to banks, insurance companies, and other major lenders in order to raise the funds necessary to purchase their steam locomotives. Since ALCo, Baldwin, and Lima custom built these locomotives to the unique specifications of particular railroads, their products could not be transferred easily to other railroads in the event of default. As such, lenders usually required an initial down payment—something they would not demand for purchases of standardized diesel locomotives.[8]

Builders did sell spare parts, which they cast or machined from wooden masters kept in a pattern vault. The multiplicity of steam locomotive designs ensured that many spare parts could not be kept in inventory, ready for immediate delivery. Additionally, since steam locomotives employed loose manufacturing tolerances, spare parts rarely fitted exactly. Once in railroad service, steam locomotives literally shook themselves apart, and the constant abrasion of metal parts against metal parts ensured that, even if the locomotive builders had achieved complete interchangeability in production, spare parts would never exactly fit the locomotive for which they were designed. Although some components, like headlights, bells, and cab assemblies, were essentially interchangeable, railroads still required highly skilled, highly paid, and effectively unionized railroad shop forces to fit spare parts to disabled steam locomotives.

Finally, since railroad employees understood steam locomotive operating and maintenance techniques, the steam locomotive builders had little incentive to offer training programs for railroad engineers or shop forces. At most, the builders supplied a representative who would ride with the engine crew of a new steam locomotive in order to evaluate its performance. As John Brown's study of the Baldwin Locomotive Works points out, *compared to other firms in the producer-oriented capital goods industry,* builders such as Baldwin offered a surprising array of marketing services.[9] Compared to consumer-oriented firms such as General Motors, however, marketing capabilities in the steam locomotive industry were scantily developed at best.

Success in the steam locomotive industry required close cooperation with railroad motive power officials, a management and a workforce willing to weather periodic peaks and troughs in locomotive demand, and managers adept at using standard components to create custom designs. As a result, steam locomotive producers possessed excellent custom-manufacturing capabilities, but less well developed managerial and marketing skills. Customers in the railroad industry contributed substantially to the success of the locomotive builders, since they occupied a central role in the locomotive design process and, therefore, controlled the pace and direction of technological innovation.[10]

Fluctuations in economic cycles affected the steam locomotive industry even more than the rest of the American capital goods industry. In general, periods of economic prosperity would tax production capacity to the limit, only to be followed by a slump in orders with attendant layoffs, plant shutdowns, and declining profits. Executives at ALCo, Baldwin, and Lima were accustomed to the "feast or famine" characteristics of the locomotive industry. Their collective attitude, essentially one of patient resignation, did not equate slumps in demand with the technological obsolescence of their products. A year in which a builder sold few or even no locomotives was not necessarily a cause for panic, or for radical changes in management or product lines, particularly if that slump in demand occurred during a period of economic depression.

During the 1920s, the three steam locomotive builders enjoyed sustained prosperity. Orders flowed in, and profits and dividends mounted. Domestic orders increased from a post–World War I low of 214 in 1919 to 1,998 in 1920. Orders peaked at 2,600 in 1922, then slowly declined until they reached 1,301 in 1926. This slow decline was followed by a sharp decrease in orders, to 734 in 1927 and 603 a year later. One last surge in demand occurred in 1929, when American railroads ordered 1,212 locomotives. During the 1920s export sales averaged approximately 220 locomotives per year.[11] ALCo, Baldwin, and Lima were building the world's most advanced and powerful locomotives, the largest pieces of machinery yet built with the capability of moving on land. Still, although their executives remained unaware of the situation, the success of these three companies was fragile. They could not foresee the economic crisis of the 1930s, nor did they understand that by the late 1920s steam locomotives had reached a technological dead end.

One way to increase the power of a steam locomotive was simply to make it bigger, but therein lay the problem. Bigger also meant longer, wider, and heavier, and by the 1920s steam locomotives had reached the capacity of railroads to support them. Track gauge was fixed and so too, for financial reasons, were the strength of bridges and the clearance of tunnels and trackside buildings. As it was, the adoption of a new class of motive power by a railroad generally required substantial additional expense for betterments to the physical plant. Multiple engines could be used for heavy trains, but each locomotive required a separate engine crew, who could communicate with their counterparts on the rest of the train only by hand and whistle signals. Thus, by the mid-1920s locomotive designers could make only marginal improvements to steam locomotive performance, utilizing devices such as roller bearings, poppet valves, streamlining, and lightweight steel alloys for valve gear and other critical parts.

These marginal improvements did not address the basic problem associated with the steam locomotive, that of thermal efficiency. Diesels ultimately replaced steam locomotives, not because they were more powerful (even today, few single-unit diesels can match the horsepower of a 1940s state-of-the-art steam locomotive), but because they can do more useful work with less fuel. As most scientists and technicians understood, even during the early years of the twentieth century, diesel engines have approximately three times the potential thermal efficiency of steam power. This inescapable constraint ultimately doomed steam power on American railroads, and ensured that marginal improvements to steam locomotive technology merely prolonged the inevitable.[12]

The Emergence of Diesel Engine Technology

The invention of the diesel engine predated its first viable application in railroad locomotives by more than a quarter of a century and attained widespread use only several decades after that. As is often the case with new technology, especially that which challenges the dominance of complex and long-established technological systems, the mere presence of the diesel engine was not enough to ensure its use. The railroad diesel engine gained widespread acceptance only after related advances in fuels, metallurgy, welding techniques, and electrical design had occurred. Ultimately, however, diesel locomotives did not begin to emerge as a viable option until improvements to existing steam locomotive technology no longer produced significant increases in performance capabilities relative to cost.

It was not until the late 1800s that the diesel engine began to assume its modern form, largely through the efforts of Dr. Rudolf Diesel.[13] Diesel's theoretical studies demonstrated that compression-ignition engines could achieve thermal efficiencies as high as 73 percent, compared to 6–10 percent for a steam engine and 18–22 percent for gasoline spark-ignition (Otto) engines.[14] In 1892, Rudolf Diesel patented his still-theoretical engine design.[15] A year later, Diesel built a one-cylinder oil-fired engine that never ran; and, by 1897, Diesel had developed an engine that produced an encouraging 26 percent efficiency when it actually operated.[16] Ironically, given his engine's ultimate success in displacing coal-burning steam locomotives from the world's railroads, Diesel initially intended his engine to burn pulverized coal, rather than fuel oil. Diesel supported the application of his engine to railroad propulsion, and in the summer of 1912, the world's first diesel-powered locomotive made its largely unsuccessful debut on the Winterthur-Romanshorn Railroad in Switzerland.

In 1912, Diesel's original patents expired, leaving the field open for other entrants. More than one hundred companies entered the diesel engine field during and immediately after World War I in response to the easing of patent restrictions and to widespread claims from Diesel and others that his engine would replace all other forms of motive power, including steam and gasoline engines. Most firms fared badly, lacking both capital and proper research and development facilities. The brief tenure of most diesel engine producers convinced many in the railroad industry that diesels, at best, would never be suitable for more than a very few specialized niche markets—in tunnels, for example, or in areas where steam locomotives posed a fire hazard.[17]

By the early 1920s, the success of diesel engines was evident only in marine applications, where weight was not an important consideration. Established locomotive builders, comfortable with steam locomotive technology, correctly assumed that they could not build a successful diesel locomotive solely by using existing technology. Furthermore, they had no interest in undertaking the research and development necessary to expand this technological envelope.

The Advantages of Diesels

Despite these promising developments, it was not until the limitations of steam power became apparent that American railroads were willing to embrace the concept of dieselization. Faced with larger and heavier locomotives and the need for increasingly expensive maintenance and betterment programs as these behemoths tore apart the physical plant, railroads considered widespread electrification programs during the early twentieth century, but the enormous expense of such a program restricted electrics to long tunnels and congested urban areas.

During the 1920s, some railroad executives expressed interest in a rapidly accumulating body of data that suggested that the use of diesel locomotives might produce substantial savings over the costs of steam locomotives without the capital requirements associated with electrification. They recognized that steam locomotive technology constituted a reverse salient in their operations, and sought methods of resolving this problem.[18] As early as 1922, one railroad official, Julius Kruttschnitt of the Southern Pacific, inquired whether Baldwin, its traditional steam locomotive supplier, might be able to build a diesel freight locomotive. Baldwin declined, claiming that all of its engineers and draftsmen were too busy designing new steam locomotives. Three years later, Kruttschnitt, in a paper delivered to the American Society of Mechanical Engineers, predicted that the replacement of steam locomotives by diesels would cut fuel costs by two-thirds.[19]

Other railroad officials echoed Kruttschnitt's interest in diesel locomotives. In January 1923, at an address to the New England Railway Club,

L. G. Coleman, assistant general manager of the Boston and Maine, complained that steam locomotives were inherently inefficient and called for further research into the possibilities of diesel locomotives. An August 1925 editorial in *Railway Age*, the leading industry trade journal, discussed "the outstanding advantages of this new form of motive power."[20] And, in May 1930, a report by the Committee on Diesel Locomotives at the International Railway Fuel Association Convention in Chicago showed that the use of diesels in switching service would lower operating costs from $7.70 to $5.09 per locomotive hour. General Electric later admitted that railroads demanded diesel transfer locomotives "even before the manufacturers were in a position to supply units of sufficient capacity."[21] These advantages remained largely theoretical during the 1920s, however, and it was not until the 1930s that technological advances in diesel engine design unlocked this potential.

Diesels had numerous advantages compared to steam locomotives. First, because of their higher thermal efficiency, diesels required less fuel to do the same useful work. By the mid-1930s, railroad executives and industry analysts understood that diesel switchers could reduce fuel costs by 75 percent.[22] These costs not only involved the expense of purchasing fuel for steam locomotives, but also that of transporting it to the fueling facilities. The transportation of coal, of which the railroad industry was the largest industrial consumer, required vast fleets of hopper cars. One railroad official estimated that the elimination of this expense alone would be more than enough to pay for new diesel service facilities. Of course, diesels also required fuel, but in typical service one tank car of diesel fuel was equivalent to eight cars of coal. In addition, diesels could travel much greater distances without refueling.[23]

Second, unlike diesels, steam locomotives required vast quantities of water. This was often difficult to obtain in arid regions of the West. Wherever it came from, water often required demineralization before it could be used in locomotive boilers. Failure to do so could result in serious damage to the boiler and, in extreme cases, explosions. In 1937, the Association of American Railroads estimated that American railroads spent fifty million dollars per year on water provision and treatment for steam locomotives. The Santa Fe, for example, used tank cars to haul more than one million gallons of water per day to the arid station at Ash Fork, Arizona. Not surprisingly, the Santa Fe was the first railroad to purchase large quantities of diesel freight locomotives. In addition, since steam locomotives could seldom travel more than a hundred miles without filling their water tank, water stops increased both travel time and railroad congestion.[24]

Third, steam locomotives had higher repair costs than diesels. Because of their weight, high boiler pressures, enormous heat production, and large numbers of loosely fitting moving parts, steam locomotives required inspection and rebuilding at regular intervals. Every year, railroads spent approximately 25 percent of a steam locomotive's original cost on its maintenance. Various estimates showed that the use of diesels would reduce repair costs 50–90 percent.[25] Since they were often out of service while building up steam, dumping ashes, or being cleaned or repaired, steam locomotives were available for revenue-generating service only 50–70 percent of the time. Railroads were typically forced to buy two locomotives for every one in constant service, a practice that substantially increased capital expenditures. Diesels, on the other hand, were often available for service as much as 95 percent of the time. In addition, since diesels needed less frequent servicing, they halved transit time, lengthened typical runs from 125 miles to 500 miles, and improved effective freight car utilization by a third.[26]

Fourth, although steam locomotives usually exceeded diesels in brute horsepower, diesels had more tractive effort, especially at low initial speeds.[27] When a diesel locomotive starts to move a train, the diesel-powered electric motors that turn the wheels operate at a low rpm setting, producing high torque, and thus high tractive effort. As a diesel locomotive accelerates, tractive effort decreases, but the momentum of the train compensates for this reduction. Furthermore, the horsepower of a steam locomotive increased as its speed increased, while that of a diesel locomotive remained constant at all speeds. These characteristics made the diesel especially suitable for switching service. At ten mph, a 600-hp diesel switcher actually had more tractive effort than a 2,000-hp steam locomotive. A 6,000-hp steam locomotive could generate 135,000 pounds of tractive effort, while a diesel of identical horsepower had an output of 230,000 pounds in low speed switching operations.[28]

Diesels offered a further advantage that no steam locomotive could emulate. Dynamic brakes, when used on trains traveling downgrade, caused the enormous inertia of the train to turn the diesel's traction motors into resistance generators. The resulting heat vented through grids on the locomotive's roof. This feature substantially reduced brake shoe wear. The Santa Fe estimated that the use of steam locomotives forced it to spend $1 million per year on brakeshoe replacement. In addition, the lack of dynamic brakes forced steam-powered trains to stop at the head of a grade so that the crew could laboriously set the retaining valves (i.e., partly apply the brakes) along the length of the train. Often, engine crews had to make further stops while descending the grade in order to cool the train's wheels. Diesels decreased traffic congestion in mountainous terrain to such an extent that many railroads shelved plans to install extensive double track after they acquired diesels. Steam locomotives filled tunnels with smoke (that occasionally proved fatal to engine crews); diesels did not. In some cases—the Cascade Tunnel, for example—railroads used electric locomotives in lieu of steam power, while diesel-powered trains could run through without stopping to change motive power.[29]

Diesels offered a variety of other important advantages. The rhythmic pounding of steam locomotive drivers, particularly when operating at high speed (a process known as dynamic augment), caused substantial damage to track and bridges and, as a result, dieselization allowed many railroads to reduce expenditures for capital improvements and routine maintenance to their physical plants.[30] Diesels offered better visibility for train crews and greater protection in case of an accident. They produced smoother starts and gentler stops than steam locomotives and produced far less pollution, all characteristics which made them highly suitable for passenger service. They were capable of higher sustained speeds and produced less of a fire hazard.[31]

Finally, diesels afforded at least the theoretical opportunity to reduce the high labor costs associated with steam locomotive operation. Even the most lowly steam-powered train usually required a conductor, an engineer, and at least one fireman and two brakemen, all of whom were well paid and well represented by powerful unions. The use of a second locomotive (a practice referred to as doubleheading) required additional locomotive crews. Since diesel locomotives had no fire, there seemed little need for a fireman; and, during the early 1930s, some builder advertisements stressed this idea. The multiple-unit capabilities of diesel locomotives allowed any number of units to be operated by only one crew, something that was not possible with steam power. Furthermore, since diesels moved trains more efficiently than did steam locomotives, railroads were able to reduce their labor costs per ton-mile. Some builder advertisements and correspondence hinted (but did not explicitly state) that post-sales support programs, such as spare-parts warehouses and rebuilding facilities, would allow railroads to reduce the size of their unionized shop forces.

The intransigence of the railroad brotherhoods offset some of the potential cost savings of diesels, however. A 1937 agreement between railroads and the Brotherhood of Locomotive Engineers and the Brotherhood of Locomotive Firemen stipulated that firemen be retained on all locomotives weighing more than 90,000 pounds.[32] During the 1940s and 1950s, General Electric produced "44-tonners," small diesel locomotives that did not require firemen because they weighed slightly less than that weight. Although the brotherhoods were willing to allow multiple-unit diesel operation with one crew, many railroads took the precaution of ordering cabless booster units to be used in conjunction with cab-equipped diesels. Some lines, such as the Santa Fe, assigned a single road number to a four-unit diesel locomotive in an attempt to forestall union insistence that the unit actually constituted four separate locomotives, each of which would require its own crew.

STEAM VS. DIESEL 19

Still, there could be little doubt that reduced labor costs contributed greatly to the overall operational savings produced by diesels.[33]

In addition to their actual and potential advantages, diesels embodied several disadvantages. These included their higher initial cost (two or three times the cost of a steam locomotive, per horsepower), their shorter life expectancy, and their need for new fuel and maintenance facilities. The cost of providing facilities dedicated to diesel locomotive repair caused many railroads to delay experimentation with diesel locomotives, especially during the Great Depression. In addition, few railroad employees knew how to operate or repair diesel locomotives. This deficiency would later give an edge to diesel locomotive producers, such as Electro-Motive, that offered comprehensive training programs to railroad operating and maintenance employees.

Technology, Production, and Marketing in the Diesel Locomotive Industry

While the building of steam locomotives constituted customized, small-batch production, diesel locomotive manufacture more closely resembled mass production. Diesel locomotives embodied three main components. These included the diesel engine itself, which in turn powered the locomotive's electrical equipment—a generator and a set of traction motors—that provided the power to turn the wheels, and finally the locomotive body, which consisted of a frame, superstructure, and trucks. Two of the components, the engine and the electrical equipment, required manufacturing tolerances as close as 1/10,000th of an inch, compared to 1/100th of an inch for steam locomotive production. These tolerances required high research and development and tooling costs which could only be amortized through long production runs. A typical diesel contained roughly ten times the number of parts as a steam locomotive, and this situation reinforced the need for rigid standardization. Thus, standardization of components *and* designs became a prerequisite to cost-effective diesel locomotive production. While Baldwin standardized locomotive components, it still offered nearly five hundred steam locomotive models. Electro-Motive had standardized locomotive designs by 1936 and in a typical year, 1950, offered only seven basic diesel models. While ALCo, Baldwin, or Lima seldom produced more than twenty locomotives to the same design, Electro-Motive averaged approximately 1,120 diesel units per major design change.[34]

Railroad motive power officials exercised little control over diesel locomotive designs, in part because they had little knowledge of the relevant technology. More important, successful diesel locomotive producers, particularly Electro-Motive, made it quite clear that they would furnish diesels, with their attendant cost savings, only if railroads agreed to accept standardized designs. Where motive power officials proved reluctant to relinquish control over the design process, Electro-Motive salesmen simply bypassed them and approached railroad finance executives instead. In particular, Electro-Motive's success depended on its ability to change the locus of design initiative in the locomotive industry from the customer to the producer.

All of the builders assigned prices to their various locomotive models, although the range of extra-cost features meant that railroads generally solicited bids for locomotive orders under the terms of the Clayton Act, as they had done in the steam locomotive industry. Although the Clayton Act mandated the solicitation of competitive bids for locomotive orders, it did not require railroads to accept the lowest available bid. In 1944, for example, the Pennsylvania Railroad expressed such a strong interest in Electro-Motive diesels that it specified the following purchase procedure:

> 1. Obtain authority from the Board of Directors to purchase a Diesel passenger locomotive.
> 2. Advertise in accordance with the provisions of the Clayton Act for such a locomotive.
> 3. Place an order, after receipt of bids, with General Motors.[35]

Clearly, many railroads did not take the provisions of the Clayton Act too seriously.

In order to preserve scarce capital, railroads generally preferred to finance diesel locomotive purchases through equipment trust certificates, as they had with steam locomotives. Lenders often negotiated more favorable equipment trust financing arrangements for diesel locomotives than for steam engines, since, in the event of railroad default, standard-design diesel locomotives could be sold to other railroads, while custom-built steam locomotives would be less likely to find a purchaser.[36]

Diesel locomotive production demanded a skilled labor force, but these skills differed from those employed in the steam locomotive industry. Welding replaced casting, knowledge of electricity superseded knowledge of steam, and theoretical knowledge supplanted practical experience. While workers built steam locomotives, they assembled diesels. At modern facilities, like Electro-Motive's La Grange plant, workers completed diesel locomotive subassemblies in bays alongside the main erecting floor, rather than at separate, isolated buildings. The steady stream of standard parts flowing into the final assembly area led many observers to talk of "assembly-lines," especially at La Grange. In reality, assembly-line manufacturing never characterized the diesel locomotive industry, since production runs seldom exceeded a thousand units, and since the two-hundred-ton-plus weight of a typical diesel generally precluded the use of conveyor belts and other accoutrements of the assembly-line system. Still, the diesel locomotive industry much more closely resembled assembly-line mass production than did the steam locomotive industry, a circumstance that undoubtedly provided a competitive edge to companies, like GM, that were familiar with standardized mass production.

Producer control over diesel locomotive design did not eliminate the importance of the customer; on the contrary, it mandated even more dedication to customer service. Railroad operating and maintenance personnel were unfamiliar with, and often deeply suspicious of, diesel locomotive technology. As such, builders found it necessary to train railroad employees in proper diesel locomotive operating and maintenance techniques. Fears that diesels would not operate as promised caused builders to both guarantee the performance of locomotives and offer warranties of up to two years. Standardization of locomotive designs allowed more rapid and comprehensive spare parts services. These services helped railroads to reduce their shop costs and gave locomotive builders the opportunity to further amortize their research and development expenditures.

Diesel locomotive production also demanded enhanced managerial capabilities. For companies like Electro-Motive, even the solicitation of locomotive orders constituted a major undertaking, since this process involved breaking the customer-driven demand patterns inherent in the steam locomotive industry. Since diesel locomotive producers assumed responsibility for locomotive designs, their managers had to establish and direct sophisticated research and development programs. The end product of this R and D was typically a sample, or "demonstrator" locomotive which toured the country's railroads. Should the demonstrator fail to meet performance expectations, or engender only lukewarm responses from railroad officials, managers could only blame themselves, not their customers. Finally, given the revolutionary nature of diesel locomotive technology, companies in that industry benefited from the presence of managers who not only understood diesel locomotive technology but who also *believed* in it, managers who had no lingering attachments to steam locomotive technology or its concomitant production requirements, managers whose primary goal was the complete dieselization of American railroads.

Performance Characteristics and Market Value

Diesel locomotives, which had virtually nothing in common with their steam-powered counterparts, clearly represented a radical technological discontinuity. Beyond this, diesels introduced an entirely new set of performance characteristics to railroad-industry customers; and executives at ALCo, Baldwin, and Lima had difficulty in even establishing the correct priorities for their marketing efforts. As builder advertisements in *Railway*

Age and other railroad industry trade journals made abundantly clear, steam locomotive producers thought in terms of horsepower. If more power could be crammed into a single steam locomotive, then so much the better. Since railroad executives disliked the "doubleheading" of steam locomotives (because of communications difficulties, the need for two separate locomotive crews, etc.), they responded favorably to large, high-horsepower steam locomotives, even when those locomotives shook their physical plant to pieces. Steam locomotive builders also advertised the undeniable fact that a steam locomotive cost only one-third as much, per horsepower, as a diesel.

Electro-Motive adopted a far different approach, one that recognized the different performance characteristics of diesels. Its advertisements stressed that the advantages of the diesel lay in operating expense reductions, not in initial cost. Since diesels could repay their purchase price in as little as three years (an impressive 33 percent annual return on investment), price was of little consequence. While diesels could not outpull steam locomotives, they had far more flexibility, since any number of low-horsepower diesels could be coupled together and operated easily by one crew. As such, particularly during the 1930s, steam and diesel locomotive builders were largely talking past each other—but railroad customers were increasingly listening to the latter and ignoring the former.[37]

II

Internal-Combustion Railcars: Springboard to Participation in the Diesel Locomotive Industry

THE SELF-PROPELLED railcar, rather than the large diesel locomotive, provided the first opportunity for the internal-combustion engine to prove itself in railroad service in the United States.[1] Railcars, similar in external appearance to conventional railroad passenger equipment, generally contained an engine compartment, a control stand, and passenger and baggage compartments. These units were entirely self-contained (unlike electric streetcars or interurbans) and so could operate even over remote, lightly traveled branch lines. Railroad interest in railcar technology occurred in two distinct phases: the first peaked shortly before World War I, and faded as more pressing wartime production and transportation needs took precedence; the second gained momentum during the mid-1920s, but was largely extinguished by market saturation and by the economic crisis of the 1930s. General Electric was the first major manufacturer to enter the railcar field, during the first "boom" in railcar demand. The Westinghouse Electric and Manufacturing Company joined the fray approximately a decade later, in response to the second wave of interest in railcar technology. Both companies were particularly anxious to exploit economies of scope by transferring their knowledge of streetcar, interurban, and straight electric railway traction to this promising new field.[2] A third player, the Electro-Motive Company (EMC), also participated in the second phase of the railcar market, although that company had a different strategy altogether. As a small start-up firm, EMC had scant knowledge of electrical equipment technology and thus little ability to take advantage of economies of scope. Instead, EMC relied on its marketing expertise to attain dominance in the railcar industry.

As railcar demand began to decline, first as a result of war, then as a result of depression and market saturation, the three companies sought to boost sales by designing and manufacturing diesel locomotives. At GE, diesel locomotive R and D efforts occurred during the late 1910s, when diesel locomotive technology was as yet too primitive. GE's unfortunate early experience in the diesel locomotive industry, like its participation in the railcar industry, had long-term benefits, however, since several GE technicians took their knowledge of electrical equipment technology to other firms, particularly EMC. And, as the second railcar boom neared its end, EMC, far more than Westinghouse, made a full-scale transition from railcars to diesel locomotives.

The Origins of Railcar Technology

The earliest railcars used gasoline engines. In 1897, the Chicago-based Patton Motor Car Company introduced a gasoline-and-battery-powered railcar, probably the first in the United States to use an electric transmission. Patton built nine similar cars between 1888 and 1893, but none of these was a great success. In 1905, the Union Pacific Railroad assigned its superintendent of motive power and machinery, William R. McKeen Jr., the task of building a gasoline-powered railcar. McKeen, as an employee of the Union Pacific and, later, on behalf of his own company, directed the production of more than two hundred railcars.[3]

The McKeen Company and its smaller competitors enjoyed scant success in the railcar market because their engines were too heavy and unreliable for railroad service. More important, no company had perfected a method for the effective transmission of power from the engine to the wheels. Most railcars, like those produced by McKeen, utilized mechanical transmissions—a series of gears that reduced the high speed of the engine driveshaft to the more sedate pace of the railcar's wheels. Mechanical transmissions were unreliable, difficult to control, and subject to frequent catastrophic breakdowns. Electrical power transmission technology offered a promising alternative, but this was beyond the limits of McKeen's organizational capabilities.

General Electric and the Railcar Industry

General Electric entered the railcar market early in the twentieth century. Whereas railroads such as the Union Pacific constructed railcars as a means of lowering operating expenses, GE saw the railcar field as a logical extension of its railway electrical equipment line. GE built its first straight electric locomotive (for the Baltimore and Ohio) in 1895 and furnished a variety of components for electric streetcars and interurbans. GE, like competitor Westinghouse, had hoped that American railroads would undertake widespread mainline electrification in the early years of the twentieth century. Primarily because of the enormous capital expenditures involved, however, electrification was generally restricted to long tunnels, underground sta-

INTERNAL-COMBUSTION RAILCARS   25

tions, and other areas where the smoke from steam locomotives created an operating hazard.[4]

When electric locomotive demand failed to materialize to the extent that had been expected, GE searched for additional ways to utilize the organizational capabilities created to produce electric locomotives. GE naturally equipped its railcars with its own electrical equipment, much of which was identical with that used on streetcars and interurbans. In order to develop a reliable power source without having to depend on outside contractors, GE established a Gasoline Engine Department in 1904. In February 1906, GE built its first motorcar, using a car body provided by ALCo. In 1911, GE established a diesel engine research and development program in Erie, Pennsylvania. By 1917, GE had produced an experimental diesel-electric locomotive, the first built in the United States. GE built three more diesel locomotives in 1917 and 1918, and all three, like the original prototype, failed to perform adequately.[5] Both these technical problems and an increase in wartime demand for more traditional products persuaded GE to terminate research on gasoline-electric and diesel-electric railcars and locomotives for the next decade. The company ended production of gasoline and diesel engines for railroad use in 1919, but continued to manufacture railroad electrical equipment. By this time, GE had lost $1.5 million on its railcar program.[6]

GE's limited railcar production (eighty-eight units between 1906 and 1914) had greater significance than the small output would indicate, since the railcar industry as a whole benefited from GE's experiments with gasoline and diesel engines.[7] Much of GE's basic research eventually found more sophisticated applications at several producers of railcars, small gasoline and diesel-powered switching locomotives, and large freight and passenger diesels. GE's willingness to install gasoline and diesel engines in railroad equipment gave further credence to that form of motive power. Finally, GE's brief railcar research and development program instigated a fruitful diffusion of railcar technology. When Hermann Lemp, a GE electrical engineer, developed a reliable direct-current electrical control mechanism, he not only overcame a serious reverse salient in railcar technology, he also established the basic design for virtually all later diesel locomotive control systems.[8] GE did not patent any important component of their railcar engines or related electrical equipment (presumably because the company had little interest in a money-losing product), and this allowed Electro-Motive to eventually produce electrical equipment that closely mimicked GE designs.[9]

In addition, many of the employees who left the GE railcar program took their knowledge and their passion for internal combustion to other companies, such as Electro-Motive. One of these GE-trained electrical-equipment experts, Richard Dilworth, joined GE as a "machinist-electrician" in 1910. GE placed him in charge of its railcar demonstration staff in 1911, and he became a test engineer the following year. In 1913, he began working on GE's diesel engine project. Although he left GE a few years after railcar development ended, he remained a staunch advocate of diesel railway traction and, as an employee of Electro-Motive, had an enormous impact on the development of the diesel-electric railroad locomotive. He served as Electro-Motive's chief engineer from 1926 to 1948, at which time he became engineering assistant to the vice president in charge of EMD.[10]

GE Enters the Diesel Locomotive Market

Although railcars proved a disappointment to GE, government legislation encouraged the company to transfer its knowledge of electrical equipment technology to the potentially more lucrative locomotive industry. During the early decades of the twentieth century, railroads serving New York City faced rising traffic levels and widespread public outrage over accidents in smoke-filled tunnels. By the early 1920s, railroads had electrified most main lines into New York City, but this capital-intensive option was simply not viable on more lightly used switching lines.[11] Steam locomotives thus continued to operate in many areas of the city. In 1923, however, the New York state legislature passed the Kaufman Act (amended in 1924 and 1926), banning steam locomotives from the entire city of New York. In June 1929, Baltimore enacted its own version of the Kaufman Act, in the form of city ordinances 746–748, which restricted or eliminated steam locomotive operation on most trackage within city limits.[12] Because it encouraged the development of experimental locomotive models, the New York and Baltimore legislation certainly influenced the timing, if not the ultimate direction and structure of the locomotive industry, and the legislation thus offers an example of the effects of government policy on technological development.[13]

The Kaufman Act provided a small initial market for diesel locomotives. Railroads that served the New York area, particularly the New York Central, approached the Ingersoll-Rand Company, an established producer of diesel engines, with a request to build a prototype diesel switching locomotive. The criteria for this locomotive, in descending order of importance, were: reliability, high potential speed, low maintenance costs, minimal noise and smoke, good fuel economy, and "reasonable first cost."[14] This set of performance criteria differed greatly from those assigned to a typical steam locomotive, since cost was at the bottom of the list, and horsepower was not even mentioned. In other words, while ALCo and Baldwin understood that their customers would grudgingly accept these performance characteristics when legally obligated to do so, railroads were not likely to buy diesels in a situation where cost and power were the sole considerations.

Beginning in 1903, GE had worked closely with ALCo to build straight electric locomotives for the New York Central's urban electrification program.[15] In 1921, General Electric built on its knowledge of electrical equipment technology in the railcar, interurban, and straight electric locomotive markets by signing an agreement with the Ingersoll-Rand Company to develop jointly an experimental 300-hp diesel switching locomotive for the NYC. This locomotive, completed in December 1923, gave its first public demonstration two months later. The following year, ALCo became a part of the production consortium by building the underframe and body of a second experimental diesel locomotive. ALCo simply served as an outside supplier of locomotive bodies, with little role in the overall design process—a situation that belies ALCo's later claims that its activities during the 1920s made the company a pioneer in the diesel locomotive industry. Ingersoll-Rand built the diesel engine at its Phillipsburg, New Jersey, plant while GE supplied the traction motors, generator, and related electrical equipment, and assembled the various components at its Erie works. Ingersoll-Rand assumed all responsibility for marketing the locomotives.[16]

After it delivered the first GE-IR-ALCo diesel switcher in July 1925, the consortium built a variety of other locomotives, all of them largely experimental, during the following years. The companies completed a 600-hp passenger locomotive in 1927, followed by a 750-hp road freight diesel in 1928. In all, the GE-IR-ALCo consortium built thirty-three diesel locomotives between 1925 and 1931. All were intended for specialized niche markets where steam locomotives, the preferred form of motive power, could not be economically or safely employed. Since the diesel engines that powered these locomotives were both overweight and underpowered, none were technologically advanced enough to threaten the dominance of large mainline steam locomotives.[17]

Because GE had developed considerable experience in carbody construction during its short pre–World War I involvement in the railcar industry, the company chose in 1927 to begin production of its own locomotive bodies. The completion of the first locomotive shells in 1928 marked the end of ALCo's involvement in this early production consortium. GE continued to buy diesel engines from Ingersoll-Rand, as well as from other manufacturers, such as Cooper-Bessemer and Caterpillar.[18] In 1929, in order to enhance its organizational capabilities in the diesel locomotive market, GE reorganized its Railway Engineering Department as the Transportation Engineering Department, later to become the Transportation Equipment Division. GE anticipated that this new department would work closely with the company's Airbrake Department and Industrial Locomotive Engineering Department, both of which were located in Erie.[19]

During the 1930s, GE produced a few dozen small switching locomotives, mostly intended for industrial and shortline railroad customers. GE introduced the "44-tonner," in 1940, and the company eventually sold more than three hundred of these locomotives. This switcher, powered by two 190-hp engines, filled an important niche market on railroads possessing weak bridges or poorly maintained track. In addition, this unit was just light enough to avoid the use of a fireman, mandatory on all locomotives weighing more than 90,000 pounds. GE used both Cooper-Bessemer and Caterpillar engines to power these locomotives; naturally, it employed GE electrical equipment.[20]

At the same time, GE continued to supply electrical equipment for ALCo's diesel locomotives. GE was careful not to produce locomotives that could compete directly with those manufactured by ALCo, an unofficial arrangement that evolved into a formal manufacturing alliance in 1940. GE experimented with a variety of other locomotives, including a 1,800-hp transfer locomotive and a 5,000-hp steam electric.[21] Still, GE concentrated on diesel switchers, since its executives assumed that "most of the heaviest hauling [on typical railroad lines] will in all probability be handled by steam locomotives for many years to come. . . ."[22]

Westinghouse Follows GE's Lead

The mid-1920s witnessed a resurgence of interest in railcar technology. Railroad companies, in response to growing automobile traffic, sought ways to reduce their expenses by substituting railcars for locomotive-hauled passenger trains in branch-line passenger service. Improved technology eliminated many of the performance limitations that had earlier mitigated against widespread railcar application. By the end of the decade, railcars had become so powerful and reliable that several railroads were using railcars to pull short trains of passenger and even freight cars, despite manufacturers' recommendations. This renewed interest in railcars drew both Westinghouse and EMC into the market.[23]

Like GE, Westinghouse used its knowledge of electric railways to advantage in the diesel locomotive industry, although the latter firm was less interested in exploiting the new niche market for diesels created by the Kaufman Act. In the 1880s, Westinghouse began to provide electrical equipment for streetcars and interurbans. Beginning in 1895, Westinghouse and Baldwin manufactured jointly mainline straight electric locomotives, with the former company providing the electrical equipment and the latter building the locomotive bodies according to specifications provided by Westinghouse. Samuel Vauclain's position as both president of Baldwin and as a board member at Westinghouse reinforced this collaboration.[24]

INTERNAL-COMBUSTION RAILCARS 29

Many of these straight electric locomotives were destined for the Pennsylvania Railroad, a loyal customer of Baldwin. The PRR first employed electric locomotives early in the twentieth century, in conjunction with its construction of Penn Station and tunnels under the Hudson River. By the late 1930s, the railroad's widespread electrification program would allow electric locomotives to operate from Harrisburg, Pennsylvania, through Philadelphia, then south to Washington, D.C., and north to New York City. This Westinghouse-Baldwin-Pennsylvania combine was thus a significant rival to the GE-ALCo-New York Central grouping, at least in the straight-electric locomotive industry. These rival groupings of manufacturers and railroads did not extend beyond the late 1920s in the diesel locomotive industry, however.[25]

Like GE, Westinghouse sought to use identical technology for straight electrics, railcars, and diesel-electric locomotives—these units often employed identical traction motors, for example.[26] Unlike GE, Westinghouse added internal-combustion engine technology to its electrical equipment capabilities. In 1925, Westinghouse built a 250-hp gasoline engine that the J. G. Brill Company installed in a railcar.[27] In 1926, Westinghouse formed a Railway Engineering Department at its East Pittsburgh works and in that same year acquired the American production rights to the dirigible engines produced by the Glasgow firm of William Beardmore Company. Westinghouse first installed Beardmore-designed engines in railcars built for the Canadian National in September 1925.[28] Although a variety of builders furnished railcar bodies to the specifications of individual railroad customers, Westinghouse assumed all marketing responsibilities and stressed that "the entire motive power equipment is 100 per cent Westinghouse with no division of responsibility for its correct functioning."[29]

After its initial foray into railcars, Westinghouse moved into the diesel locomotive industry. The company completed a pair of 660-hp diesel switching locomotives for the Long Island Railroad in January 1928. Westinghouse assembled the locomotives' electrical equipment and "Westinghouse-Beardmore" diesel engines (the first of their type to be used in railroad service in the United States) at its South Philadelphia facility. Baldwin constructed the locomotive bodies.[30] Westinghouse also delivered locomotives to the Canadian National in 1928. While Baldwin had responsibility for the mechanical design of these export locomotives, the company had to work closely with Westinghouse, the Canadian National Railways, the Canadian Locomotive Company, and the Commonwealth Steel Company—a complicated collaboration that reduced substantially Baldwin's control over the manufacturing process.[31]

In 1929, Westinghouse and Baldwin agreed to cooperate in the production of 400-hp and 800-hp diesel switching locomotives, with the former company supplying the electrical equipment and the diesel engines. Westinghouse designed and marketed these locomotives, and Baldwin served only as an independent supplier of bodies, underframes, and running gear.[32] A year later, Westinghouse spent $300,000 on improvements to its South Philadelphia plant and consolidated all of its locomotive production at that location.[33]

By the early 1930s, Westinghouse offered a standardized line of diesel locomotives that were technologically sophisticated, at least by contemporary standards.[34] Westinghouse locomotive engines, still based on the earlier Beardmore design, ranged from a four-cylinder, 265-hp model to a twelve-cylinder, 800-hp unit. These engines were exceptionally light (sixteen pounds per horsepower), largely as the result of extensive use of aluminum alloys. While these Westinghouse diesel engines produced a cleaner exhaust than did their competitors, they experienced frequent trouble with their fuel injectors, bearings, and other mechanical parts.[35] In addition, Westinghouse was never able to manufacture these engines on a mass-production basis—a problem common to many early diesel engine designs.[36] The Westinghouse "visibility cab," introduced in 1929–30, constituted another technological refinement, one that was not yet available from any competitor. The sloped sides of this design provided excellent visibility for engine crews without exposing them to the operational hazards of GE-IR-ALCo units, which forced the operator to sit at the extreme front of the locomotive. The single, centrally located cab required only one set of control equipment, producing substantial savings over the cost of a typical dual-cab locomotive. Although Westinghouse produced only fifteen "visibility cab" units between 1929 and 1937, all other builders copied this basic concept for many of their own locomotive designs.[37]

While Westinghouse had completed fifteen railcars and thirteen locomotives by the end of 1931, the Great Depression prevented the company from taking advantage of the 1927 dissolution of the GE-IR-ALCo consortium, and locomotive sales were considerably lower than expected.[38] Between 1928 and 1936 Westinghouse built only twenty-six diesel locomotives.[39] Between August of 1930 and early 1933, Westinghouse received no locomotive orders at all. By 1934, orders were still few and far between. Westinghouse introduced new locomotive designs, but the company also produced non-standardized custom units designed to meet the performance requirements specified by particular railroads.[40]

By the early 1930s, the limited potential of the gasoline-powered railcar, daunting technological imperfections in diesel engine propulsion, and the onset of the Great Depression combined to convince executives at Westinghouse to supply electrical equipment to other companies, rather than build locomotives on its own account. Westinghouse announced in June 1936, that it would no longer take orders for diesel locomotives or the Beardmore engines that powered them. The company continued to supply electrical

INTERNAL-COMBUSTION RAILCARS   31

equipment to Baldwin and to other diesel locomotive builders, provided that railroad customers specified Westinghouse electrical equipment.[41] Westinghouse purchased substantial blocks of Baldwin stock in order to solidify this market for electrical equipment, ensuring that when Baldwin executives committed their company to extensive diesel locomotive production after 1939, they did so under the watchful eye of Westinghouse.[42]

Electro-Motive and the Railcar Industry: The Emergence of a Marketing-Based Company

While both GE and Westinghouse depended on economies of scope in the electrical equipment market, the new Electro-Motive Company relied on the marketing expertise of its founder, Harold L. Hamilton, and on Hamilton's success in gaining control over the design process.[43] Born in California in 1890, Hamilton showed an early aptitude for mechanical devices. He embarked on a career in railroading and by 1914 had advanced to the position of road foreman of engines for the Florida East Coast Railroad. In that year, he joined the White Automobile Company in Denver, working in both the engineering and sales departments. At White, Hamilton led that company's efforts to adapt its highway vehicles to railroad use. His other tasks included teaching teamsters how to operate and maintain trucks, rather than horses. These instructional duties gave Hamilton valuable marketing experience and taught him how to overcome entrenched ideas regarding the most suitable form of motive power. For a short time during World War I, Hamilton served as a member of the Engineering Committee of the Army Motor Transport Corps. By the 1920s, Hamilton had become familiar with the limitations and the potential of internal combustion engines and had no vested interest in slowing the diffusion of internal-combustion locomotive technology.[44]

Although he was moving upward in the White organization—he became western wholesale manager in 1921—Hamilton decided to cast his lot with the railcar industry. He resigned from White in mid-1922 and on August 31 of that year founded the Electro-Motive Engineering Corporation (he changed the name to the Electro-Motive Company in late 1923). Hamilton soon recruited Ernest Kuehn, Andrew Finigan, Tom Finigan, and Jimmie Hilton, four employees who had been part of GE's now-defunct railcar program. Many of these employees had worked on the GE-IR-ALCo diesel locomotive program during the early 1920s. Hamilton also hired Paul R. Turner, who had worked for White Motor Truck from 1918 to 1922. Turner joined EMC in 1922, and established EMC's New York office in 1925. He later became eastern regional manager, then director of sales.[45]

Designing and Subcontracting

Strictly speaking, EMC did not manufacture anything. Instead, company engineers and draftsmen designed railcar components and subcontracted their manufacture to other companies. EMC sales agents then marketed the finished product to customers in the railroad industry. Because railcars were similar in design, structure, and appearance to regular (unpowered) railroad passenger equipment, and because car builders were familiar with standardized designs, EMC subcontracted the production of railcar bodies to the St. Louis Car Company (which manufactured most of EMC's railcar bodies), the Pullman Company, the Osgood-Bradley Car Shops, the Standard Steel Car Company, and the Bethlehem Steel Company.[46]

Although Westinghouse sold some of its products to EMC, General Electric served as EMC's most important supplier of electrical equipment. Several highly skilled GE employees, including electrical engineer Richard Dilworth, worked closely with EMC to improve electrical equipment designs, paying particular attention to electrical power control devices.[47] Hamilton found a receptive audience among GE engineers "who still had faith in the whole [railcar] program." And, as Hamilton explained, "by working backwards through this group of individuals . . . we revived the general interest in the [railcar] program"—despite an earlier GE corporate decision to abandon railcar technology as a lost cause. In response to Hamilton's considerable persuasive talents, Dilworth eventually left GE in 1926 and for the next twenty years served as one of Electro-Motive's leading design experts.[48]

While EMC purchased electrical equipment and car bodies from a variety of manufacturers, the company remained loyal to one manufacturer, the Winton Engine Company, for its railcar engines.[49] Winton initially supplied gasoline engines for EMC railcars, thanks largely to Hamilton's ability to sell Alexander Winton on the railcar idea. Since Winton was in receivership during the early 1920s, "the management of it and the bank did not feel that this proposition offered enough future and was secure enough . . . for them to venture the capital involved to develop a new engine." Although a gasoline railcar engine R and D program seemed to offer so few financial rewards, two Winton employees, Chief Engineer Carl Salisbury and General Manager George Codrington, like Hamilton, believed strongly in railcar technology. As a result, Codrington and Hamilton made a direct appeal to Alexander Winton. The project enthralled Winton, who, "even in his last years, . . . had the spirit of the pioneer," and he agreed personally to pay the cost of the railcar engine R and D program.[50]

By 1924, the Winton Engine Company had completed the development of a 175-hp gasoline engine, and the company increased the output of its engines to 220 hp in early 1925 and to 275 hp later that year. By 1927, Winton offered 300-hp, six-cylinder, and 400-hp, eight-cylinder, gasoline engines for railcar service. Gasoline was expensive, however, and also created a potential fire hazard. In the late 1920s, EMC accordingly asked Winton to develop a new railcar power source. Primarily because EMC was its most important customer, Winton began a cooperative research project with Richard Dilworth and other EMC engineers to develop a distillate engine. Distillate, similar to kerosene, was about one-fifth as expensive as gasoline. Although Dilworth believed that Winton's distillate engine was not a great success, EMC did offer some railcars with distillate engines in the early 1930s.[51]

Diesel engines offered the best potential power source for railcars, assuming that their size and weight could be reduced to manageable proportions. Winton had produced its first diesel engine in 1913, a 175-hp stationary behemoth used to provide power for Winton's Cleveland factory. By 1916, Winton had three sizes of marine diesel engines in production. One of these, the Model W-40, weighed forty-five tons, yet produced only 450 hp (an abysmal ratio of ten horsepower per ton), making it far too heavy for railroad applications. In 1928 EMC and Winton began a cooperative research and development program to design more suitable diesel engines. These efforts failed, however, owing largely to Winton's poor production techniques and limited technical knowledge. As a result, EMC continued to use Winton gasoline and distillate engines until 1934.[52]

Marketing Innovations

EMC's real success lay in its ability to retain control over the design process. While EMC did build some railcars to meet the specific operating requirements of particular railroads, the company's products more closely approximated standardization than did steam locomotives. By controlling railcar design, EMC could insist on design standardization, which in turn produced substantial production efficiencies. The cost savings that resulted made EMC railcars more marketable; and as the company sold more standard-design railcars, it was able to amortize its research and development and design expenditures. EMC incorporated collective customer requests for alterations and improvements en masse in periodic design changes, a process similar to that employed by GM in the auto industry.[53]

Although railroad motive power officials had scant familiarity with internal-combustion technology, and could thus offer little practical contribution to the design process, they were still in a position to oppose railcar purchases. Accordingly, Hamilton and his colleagues bypassed entrenched railroad motive power officials and instead approached top management officials, particularly those in the financial departments. Far more than motive-power officials, they appreciated the cost savings associated with internal-combustion railcars. Hamilton reasoned, "We were going to sell these cars to the top management and work downward, as far as necessary, rather than up through the organization as was conventional. We were selling a product entirely on 'economy and performance,' which likewise was new and different."[54] More pragmatically, Hamilton said he and his staff "went to the president of the railroad as a rule, and told him . . . that their mechanical people and their staff people were in no position to process the buying of a thing like this through their normal procedures."[55]

In order to reassure railroad executives about purchasing a product that they had not helped to design, EMC offered training programs, guaranteed the performance of its railcars, offered warranty protection, and provided rapid parts-replacement services. While at White, Hamilton had created a service and demonstration staff that was twice the size of his sales staff, and he continued his commitment to training programs after founding EMC. Hamilton realized that steam locomotive engineers had to be taught how to operate railcars, just as teamsters had to be acclimated to the mysteries of the motor truck.[56]

Typically, an EMC field instructor traveled with every new railcar for at least thirty days. In addition to instructing engineers in the proper operating procedures, instructors also rode with the railcars in order to fix any unexpected mechanical problems and to make sure that the cars were being serviced correctly. This use of "riders" continued well into the diesel locomotive era, when increasing reliability and improved railroad training programs made their services unnecessary.[57]

H. B. Ellis, former service manager at White, joined EMC in April 1926, and largely created the company's warranty protection and spare-parts service organization. Ellis established the unit exchange system, in which a railroad turned in a defective part and received a new one immediately, without waiting for the old part to be repaired. With the unit exchange system, railroads usually received their replacement railcar parts within twenty-four hours of placing an order—by chartered plane, if necessary. EMC also established regional parts warehouses in order to expedite customer orders. Most important, even though EMC was not itself a manufacturing firm, the company assumed responsibility for all of the parts used in its railcars, no matter who produced them. By stocking and distributing parts manufactured by Winton and GE, EMC increased its revenues, enhanced customer loyalty, and reduced the likelihood that its suppliers would use their own sales and service network to begin to compete directly with EMC.[58]

Market Dominance and Market Saturation

EMC's innovative marketing tactics served the company well during the 1920s. Business increased steadily after EMC delivered its first railcar in July 1924. The company sold some five hundred railcars over the next six years. By 1930, EMC had several hundred employees, mostly in the mechanical, engineering (drafting), sales, and service departments. Between 1924 and 1930, EMC captured 84 percent of the railcar market. While EMC's railcar designs were as good as any in the industry, the company's success had more to do with its marketing abilities, especially its post-sale service and repair programs.[59]

By the end of the 1920s, EMC railcars trundled along rail lines throughout the United States. Some saw service abroad on the Mexican National, the Victoria Railways of Australia, and the Central American lines of the United Fruit Company. The Chicago, Burlington, and Quincy Railroad, which operated numerous lightly patronized midwestern branch lines, purchased more EMC railcars (fifty-seven) than did any other customer. The Santa Fe, the Chicago and Northwestern, the Rock Island Lines, the Lehigh Valley, and the Northern Pacific each owned twenty-five or more EMC railcars. Significantly, the Burlington and the Santa Fe both became strong supporters of EMC passenger and freight diesel locomotive technology during the 1930s. Altogether, forty railroads used EMC railcars in regular passenger service. This extensive market penetration gave EMC engineers experience in the application of internal-combustion technology to a wide variety of climactic conditions, load factors, and operational requirements. In addition, EMC created substantial goodwill among executives at these companies, impressing them with marketing initiatives and a comprehensive knowledge of internal-combustion technology.[60]

By 1930, however, a saturated railcar market and the onset of the Great Depression had sharply reduced EMC's sales, and it seemed to many in the railroad industry that both EMC and internal-combustion railcar technology had reached a dead end.[61] GE and Westinghouse had been able to escape the dwindling railcar markets of the late 1910s and late 1920s by designing and building diesel switching locomotives for specialized applications, but EMC lacked both the money and the technological knowledge necessary to expand into this market. During the late 1920s EMC had developed an unsuccessful diesel-electric switching locomotive prototype, but Hamilton estimated that the company would need to spend $5 million in order to resolve the many technical problems associated with the project—money that EMC did not possess.

In retrospect, the railcar experience proved to be of immense value to

Electro-Motive. As Hamilton understood, railcars "gave the railroads an appetite for the economies of internal-combustion power."[62] Railcar production allowed EMC to develop the organizational skills, especially in marketing, that enabled its successor, EMD, to dominate the diesel locomotive market in the decades that followed.[63] These skills had their ultimate origin in the automobile industry, a characteristic that made EMC's corporate culture compatible with that developed by General Motors. Unlike managers in the steam locomotive industry, EMC executives had no vested interest—no stake—in the maintenance of steam locomotive technology. They were instead receptive to the concept of internal combustion, be it in the form of gasoline or diesel engines, as replacement technology. EMC executives had nothing to lose by the extinction of steam locomotives on American railroads and much to gain. While EMC's marketing expertise produced success in the railcar industry, the company became a viable diesel locomotive producer only after it had combined its marketing skills with the considerable technological know-how provided by General Motors. This marriage of strengths, and of complementary corporate assets, allowed EMC to achieve remarkable technological and marketing, if not financial, success during the 1930s.

III

First-Mover Advantages and the Decentralized Corporation

THE DIESEL locomotive industry came of age during the 1930s. By the end of the decade diesel switchers had conclusively demonstrated their superiority over their steam-powered counterparts, and improvements in diesel engine and electrical equipment technology had unleashed the potential for widespread mainline freight dieselization. Diesels replaced steam locomotives in three main stages. By 1935 most railroads had accepted the superiority of diesels over steam locomotives in yard switching service. Passenger service next felt the effects of dieselization, ensuring that diesels powered many luxury trains by the late 1930s. When the United States went to war in December 1941, railroads were just beginning to apply diesels to road freight service. Depression-induced financial constraints prevented railroads from purchasing large numbers of diesels during the 1930s. In addition, many conservative railroads, especially those with a high percentage of coal traffic, adopted a wait-and-see attitude toward dieselization during the 1930s. As a result, steam locomotives still accounted for more than 90 percent of all locomotives in service on American railroads in 1940. Of more concern to locomotive producers, however, diesel locomotive orders exceeded those for steam locomotives nearly every year during the 1930s. Furthermore, because railroads often standardized on the models of only one builder, the company that acquired the largest market shares during the 1930s would establish a substantial first-mover advantage. That company was Electro-Motive.[1]

In 1930, ALCo and Baldwin were strong, successful companies, while EMC faced market saturation and financial catastrophe. Ten years later, EMC was the dominant producer in the locomotive industry, with ALCo and Baldwin running a poor second and third, respectively. The decade of the 1930s gave EMC the opportunity to combine its organizational strengths with the financial resources of General Motors. This marriage was initially unintentional, and it did not guarantee EMC a dominant role in the diesel locomotive industry. In spite of GM's later claims to the contrary, during the early 1930s its senior management showed little interest in the potential of diesel locomotives. GM's initial involvement in this industry occurred as much from chance as from careful corporate planning.

GM and Automotive Diesels during the 1920s

GM's involvement in the diesel engine industry grew directly from its production of automobiles. Even in the early years of the twentieth century, diesel-engine advocates realized that the largest potential market for diesels lay in the motor vehicle industry. In 1921, GM engineer Carl E. Summers began the first diesel engine tests at that company, using a model from the Schwer Engine Company of Sandusky, Ohio. Results were not encouraging, and GM soon dropped the testing program. When Alfred P. Sloan Jr. assumed the GM presidency in 1923, he placed greater emphasis on research and development and expansion into related product lines. This GM diversification program included the 1925 purchase of the Yellow Truck and Coach Company and the 1929 acquisition of the Allison Engineering Company, a firm that had done some initial research on diesel airplane and dirigible engines. In addition, GM broadened its research to include fuels and metals, knowing full well that advancements in these fields would have applications in both diesel and gasoline-powered vehicles.[2]

Packard's announcement that it had developed a diesel aircraft engine in the autumn of 1928 was enough to stir GM into action.[3] After completing a lengthy study of various types of automotive gasolines in early 1928, Charles F. Kettering, the director of GM's Research Laboratories, began to examine the feasibility of using diesel engines in trucks, buses, and automobiles.[4] Gasoline research, particularly that concerning the role of the anti-knock compound tetraethyl lead, convinced Kettering that information on gasoline engine combustion should apply equally well to diesels.[5] According to Kettering, "A study of diesel engines seemed to be a direct supplement to the work which we had been doing in connection with Ethyl gasoline."[6]

Kettering found the diesel a difficult proposition. In 1928, he wrote to a colleague, "At the present time my opinion of the diesel engine is not fit to put in print."[7] Kettering realized that the most significant of the many problems plaguing diesel engines were excessive weight and inadequate fuel injectors. Injectors, designed to spray a fine mist of diesel fuel into the cylinders under high pressure, frequently malfunctioned as a result of high temperatures, metal fatigue, and poor construction. In addition, Kettering realized that metallurgy had not yet caught up with diesel-engine technology. Because of their compression ignition, diesel engines typically operated with a cylinder pressure of 650 pounds per square inch, compared to 125 pounds for a gasoline engine using spark ignition. While gasoline engines had compression ratios of six-to-one, diesels utilized compression ratios of sixteen-to-one. Without high-strength metal alloys, engine parts would therefore need to be very heavy to be strong, and this excessive weight was clearly unacceptable.[8]

During the late 1920s, Kettering moved from theoretical research to actual experimentation with diesel engines. In March 1928, Kettering began testing a single-cylinder Cummins diesel engine. A month later, he purchased a yacht, the first *Olive K*, which he used as a floating test bed equipped with Bessemer diesel engines. This yacht, along with the Winton-powered second *Olive K* (launched in September 1929), gave Kettering valuable information that was later put to good use in the redesign of the fuel injectors.[9]

The onset of the Great Depression nearly terminated GM's brief foray in the field of diesel research. The crisis of confidence that plagued many companies after the stock market crash caused GM to suspend virtually all diesel research in November 1929. Thanks to persistent lobbying by Kettering, research resumed a month later.

Anxious to avoid a long and expensive R and D program, Sloan put pressure on Kettering to purchase an established producer of diesel engines. During the second half of 1929, GM offered to buy the Treiber Diesel Engine Company for $200,000, but a disputed contract with the Consolidated Shipbuilding Company canceled the arrangement. GM also failed in its efforts to acquire the Cummins Engine Company. GM succeeded on its third attempt, purchasing the Winton Engine Company.[10]

GM acquired Winton as a wholly owned subsidiary in June 1930, but the automaker had little interest in Winton's products. Instead, GM wanted to combine the technical expertise of GM engineers with Winton's facilities. Geography provided one of GM's main inducements for purchasing Winton, since diesel engine modifications designed by GM engineers in Detroit could be transmitted quickly to Winton technicians in Cleveland.[11] In order to reinforce GM control over its new subsidiary, Kettering sent his son Eugene to Cleveland to take charge of Winton's experimental diesel engine program.[12]

Of the many changes that Kettering and the other GM engineers made to the Winton engines, two were especially important. The first was a redesign of the fuel injectors, which incorporated elements of both the GM and the Winton research programs. Many early fuel injectors used compressed air to force the fuel into the cylinders. These compressors wasted a large portion of the engine's total horsepower, and the high-pressure fuel lines often leaked—a dangerous problem near a hot engine.[13] In 1928 and 1929, both the GM Research Laboratories and Carl Salisbury, chief engineer at Winton, were independently working on designs for a unit fuel injector. This injector "combined all of the functions of metering, pressurizing and atomizing" of the fuel and was a marked improvement over earlier systems.[14] Unfortunately, Winton simply did not have the manufacturing capabilities to produce the unit injector, which required tolerances of 1/10,000 of an inch. GM's Research Laboratories therefore manufactured the injectors until 1937, when GM transferred production to the newly created Detroit Diesel Division.[15]

The Research Labs made a second improvement to the diesel engine by replacing Winton's four-cycle design with a simpler, lighter, two-cycle design. As was the case with the unit injector, GM's contribution largely involved refining and improving designs that already existed. The four-cycle engine was so named because the piston fired only on every fourth stroke. A compression stroke was followed by a power stroke, then an exhaust stroke, and finally an intake stroke. The two-cycle engine had only a compression/exhaust stroke, followed by a power/intake stroke. Kettering believed that the use of the two-stroke engine was the most effective way to reduce excess weight, and, to a certain extent, he was correct.

GM went too far, however, in asserting that "the two-cycle lightweight diesel engine was entirely a development of General Motors research."[16] As early as 1879, Englishmen William Barnett and Sir Douglas Clerk had designed a two-cycle gasoline engine. Busch-Sulzer Brothers had already been producing two-cycle diesels for more than twenty years in both Germany and the United States. By the early 1930s, most diesel engineers were familiar with the relative merits of two-cycle and four-cycle engines, generally assuming, based on differing driveshaft speeds, that the former were best suited to large engines, and the latter to small and medium size engines.[17]

Furthermore, despite GM's claims that the two-cycle diesel was far superior to the four-cycle, both were (and remain) viable in railroad applications. Some railroads, such as the Burlington, gave preference to two-cycle diesels because they were better suited to fast lightweight passenger trains, a preference that would have important repercussions for EMC. Still, companies that used four-cycle engines, such as ALCo, Baldwin, and GE, were not inherently doomed by GM's "invention."

What did more to set GM apart from its rivals was that company's use of recently developed high-strength metal alloys, its involvement in diesel fuel research, and its ability to produce engine parts to extremely close tolerances. GM and Winton not only designed components that produced incremental improvements in diesel engine technology, but they also developed the specialty steels and testing apparatus necessary to translate engineering concepts into practice. In addition to the new unit fuel injector, GM and its subsidiary developed improved crankshaft bearings and new metal alloys for those bearings, as well as high-strength pistons and instruments to measure piston temperatures.[18]

In spite of GM's impressive advances in diesel engine technology, it found few commercial applications for its products. Although GM was primarily interested in the motor-vehicle market for diesel engines, the company was unable to develop an adequate automobile or truck engine during the 1930s. And, during the first half of the decade, GM had no interest in the railroad diesel market.

Instead, much of the impetus for continued GM diesel development came from the American military, long a sponsor of leading-edge technology. In November 1932, the Navy ordered a single twelve-cylinder GM-Winton diesel engine for consideration for possible use in submarines. By April 1933, this engine, the Model 201, was providing fairly reliable results in tests.[19] In November of that year, the Navy ordered sixteen additional diesel engines from Winton. Other orders followed, but the Navy's continued patronage depended on finding a quick solution to production problems at Winton.[20]

After acquiring Winton, Kettering and the other engineers at GM discovered, to their dismay, that their new subsidiary continued to follow its own agenda. Winton employees were used to custom fitting and frequently accepted loose tolerances between parts. GM engineers were horrified to discover that these workers frequently drilled holes in engines that were already partly assembled, leaving metal chips inside the engine. Even worse, when GM's research labs were ready to make the shift from four-cycle to lighter two-cycle engines in early 1932, Winton refused to follow its parent's lead. Winton's engineers continued to have faith in the four-cycle principle and would not fully support the two-cycle engine until early 1934. Kettering complained, "You haven't got a man in Winton that wants to make a two-cycle engine so it will run."[21] By July 1935, Kettering wrote of Winton that "as it is, they are going ahead building and selling more mistakes which will ultimately ruin our prospects," and that "the failure of Winton to [establish] an intelligent program of engineering progress affecting future models has been very disappointing."[22]

"Capital That Wouldn't Control"

By 1930, as GM was making arrangements to purchase Winton, Hamilton realized that Electro-Motive could not survive as an independent company. Since EMC had saturated the railcar market, diesel locomotives offered the only realistic possibility for further growth. Hamilton estimated, however, that it would cost EMC approximately $5 million to develop a reliable diesel switching locomotive, with an additional $5 million to tool up for quantity production. EMC did not possess those kinds of resources. Moreover, as Hamilton recalled in 1957, "the money itself would not have been enough. Winton didn't have the technological men and equipment required for development of the engine we needed. General Motors did."[23]

GM purchased Electro-Motive in December 1930, only because that company provided a captive, if small, market for Winton engines, not because GM planned a comprehensive diesel locomotive development program. EMC President Harold Hamilton later said that "after negotiations with officials of General Motors and the Winton Engine Co. had progressed toward a decision it became evident to General Motors that the Winton Engine Co.'s biggest customer was the Electro-Motive Corporation, and in those days it was almost a case of the tail wagging the dog."[24]

At the time it purchased Winton, GM had no plans to make a long-term commitment to the diesel locomotive industry, primarily because that company saw so little profit in it. To put it bluntly (as Hamilton did) EMC was "not in any profit position as an enterprise . . . at that time, to interest General Motors or anyone else."[25] Instead, GM's decision had more to do with professional relationships and personal friendships that had been established among the community of engineers and technicians at GM Research Labs, Winton, and EMC. In particular, Hamilton established a close relationship with Kettering, who in turn served as an advocate for diesel locomotive technology within the GM corporate bureaucracy.[26]

Hamilton had some very definite ideas about EMC's role in the vast GM corporate hierarchy, even if GM did not share those goals. In particular, he was determined to use GM as a source of capital but "capital that wouldn't control, so that we could still manage the business and carry the project that we were shooting for, carry it on."[27] Beyond viewing GM as a source of capital, Hamilton had little knowledge of that company's labyrinthine corporate structure, and he "did not know General Motors' organization then, how they operated, and who the people were, anything about them, except Mr. Kettering."[28] Hamilton and his associates soon discovered, to their delight, that Kettering was very much intrigued by the possibility of applying diesel engines to locomotives. As Hamilton recalled, "The minute we fed him that [idea] it was just like ringing a bell to a fire horse."[29] Kettering, who had already established a considerable reputation at GM, nurtured EMC's locomotive development program, although without fully informing top GM officials of the resources required for such a project.

With the assurance of at least a small market for maritime diesel engines, Kettering and his colleagues examined the possibility of modifying the Model 201 Navy engine for "generator sets and industrial use" but made no mention of possible railroad applications.[30] By mid-1932, GM had decided to demonstrate the industrial utility of its diesel engines to the general public by using two eight-cylinder Model 201 diesel engines to provide power for the GM building at the 1933 World's Fair in Chicago. These engines, each of which provided six hundred horsepower, performed adequately but not spectacularly. Frantic nighttime maintenance was often required in order to have at least one engine serviceable the following day, and spare parts flowed with annoying regularity from Detroit and Cleveland to Chicago. The engine performed so erratically that GM technicians lived in the basement of the GM building, ready to perform emergency repairs at a moment's notice.[31] A "Summary Report of Operation of Winton Diesel Engines at [the] Century of Progress" listed eight pages of defects and their possible solutions and indicated that something went wrong with virtually everything on the engines at least once.[32] Eugene Kettering said afterward that "the only part of that engine [sic] that worked well was the dipstick."[33]

In spite of these difficulties the general public was suitably impressed with the GM Chicago World's Fair diesel engines. Ralph Budd, president of the Chicago, Burlington, and Quincy Railroad, was especially taken by them. During the 1920s, the branch-line-encumbered Burlington had been EMC's largest railcar customer. In September 1932, Budd had begun to design a lightweight, streamlined train for the railroad's premier Chicago-to-Denver service. Hamilton, on his own initiative, approached Budd with a proposal for a diesel-powered train and took him to see the two World's Fair engines being built by Winton. Budd had originally intended for the train to be pulled by a streamlined steam or distillate locomotive but, after seeing the Winton diesels, decided that he had to have an ultramodern diesel engine pulling his ultramodern train.[34]

Budd received a cool reception from Sloan, who believed that a frantic R and D program to place a largely untested diesel engine in a highly demanding railroad application would result in ignominious disaster. Budd, however, had a powerful ally in Hamilton's deeply held faith in the widespread applicability of the diesel engine to railroad service. In addition, Budd and Kettering enjoyed a close friendship, and the railroad president was described as "an admirer of Ket."[35] One factor that may have tipped the balance was the fact that Budd "started the design and building of the train without having a prime mover to go in it, and I think it was the pressure of this outside influence which the Boss [Kettering] used very effectively with the people in General Motors."[36] Eventually, Budd, Hamilton, and Kettering were able to convince Sloan and other GM top management to provide a Model 201 diesel engine for the Burlington *Zephyr*.[37]

The train performed magnificently.[38] On its test run, on April 17, 1934, an engineer who had never before operated a diesel locomotive took the *Zephyr*, in the space of six miles, from a dead stop to 108 miles per hour. The train accelerated so fast that fuel oil sloshed up and out of its tank and the train's tail lights fell off. On its official first day of operation, the *Zephyr* traveled more than one thousand miles from Denver to Chicago, nonstop, in thirteen hours and five minutes, thirteen hours faster than the previous record pace.[39]

The *Zephyr*'s long-term benefits were even more gratifying. The train cost only thirty-one cents per mile to operate, less than half the cost of a steam locomotive and conventional passenger equipment. The *Zephyr*'s faster service caused passenger loads to increase by 50 percent. Its speed also increased the likelihood that the Burlington would receive and retain important mail contracts. The train's tremendous public appeal also produced significant benefits for EMC. On the day of its inaugural run, businesses closed and schools let out so that adults and children along the route could watch the *Zephyr*, soon nicknamed "Little Zip," flash past at more than a hundred miles per hour. Because of its speed and shiny stainless steel exterior and because it did not look at all like a steam locomotive, the *Zephyr* drew crowds wherever it went. The popular press praised the *Zephyr* as an example of what American ingenuity could accomplish, even during the dark days of the Great Depression. In 1934, RKO Pictures released *The Silver Streak*, a forgettable movie that nonetheless exposed millions of Americans to the *Zephyr*.[40]

Other railroads soon demanded passenger trains similar to the *Zephyr*. Although they could seldom afford to buy traditional steam locomotives during the depression, railroads were willing to expend considerable sums on these luxury trainsets because they reduced operating costs, increased publicity, and guaranteed capacity bookings, since the very rich were able to travel in luxury, no matter what the state of the economy. This new demand for lightweight trainsets was nothing short of salvation for EMC, which had sold only one railcar between June 1932, and the beginning of 1934. EMC coordinated the production of thirteen trainsets over the next three years. As had been the case with its production of railcars during the 1920s, EMC allocated the production of components to others. Winton supplied the diesel engines and GE and Westinghouse furnished electrical equipment, perpetuating the manufacturing arrangement that EMC had developed in the railcar industry during the 1920s. Carbuilders, such as the Budd Company and the Pullman Company, built the locomotive bodies.[41]

EMC first completed the M-10000 *City of Salina* for the Union Pacific. This three-car, largely aluminum trainset could attain 110-mph speeds. Because the Union Pacific was willing to accept a 600-hp distillate engine, EMC delivered the M-10000 in February 1934, two months before the diesel-powered *Zephyr* was ready for tests. Like the *Zephyr*, it created an immediate public relations sensation. Other diesel-powered trainsets soon joined the *City of Salina* on the Union Pacific: the five-car, diesel-powered M-10001 (*City of Portland*) in October 1934, and the M-10002 (*City of Los Angeles*) and M-10003 (*City of San Francisco*), both in service by June 1936. The Boston and Maine took delivery of a three-car *Zephyr* clone, the *Flying Yankee*, in February 1935. The Burlington received two larger *Zephyrs* for the Chicago-St.Paul run, as well as a *Mark Twain Zephyr* for service between St. Louis and Burlington, Iowa.[42]

Fortuitous technological linkages aided EMC's success in the diesel-powered trainset market during the mid-1930s. Since typical railroad passenger cars weighed approximately ninety tons apiece, a train of even modest length would have vastly exceeded the haulage capabilities of the Model 201 engine. Fortunately for EMC, two railroad car builders were independently working to reduce passenger car weights and thus unintentionally brought them within the abilities of early diesel locomotive power plants. By 1933 the Edward G. Budd Manufacturing Company, builders of the Burlington *Zephyr*, had perfected the Shotweld process. Shotwelding constituted the first practical method of joining large pieces of stainless steel, a metal that possessed an extremely high strength-to-weight ratio. Budd's extensive use of stainless steel and aluminum meant that the *Zephyr* weighed 219,000 pounds, compared to 809,000 pounds for conventional passenger equipment. Unable to acquire Budd's patented Shotweld process, the Pullman Company employed a steel alloy, Cor-Ten, that United States Steel had developed in 1934. Cor-Ten had more than twice the strength of the steel traditionally used in passenger car construction, and its use accordingly reduced the car weight. As such, the timely arrival of lightweight equipment coincided exactly with EMC's development of reliable, although underpowered, diesel-powered passenger equipment. While older heavyweight passenger cars remained on railroad rosters well into the 1950s, these new "lightweights" gave EMC the opportunity to demonstrate its technological advancements at a crucial juncture.[43]

Diesel-powered luxury trainsets had their drawbacks, however, and they would not provide a sustainable market for EMC. Since all units of the trainset were semipermanently coupled, a minor problem in one car could disable an entire train. This lack of flexibility also prevented railroads from increasing or decreasing the length of the train to accommodate changing passenger demands. The custom-built trainsets were smaller than conventional equipment and thus could not accommodate other cars. The Union Pacific's practice of assigning an electrician and a road foreman of engines to its trainsets at all times suggests that its diesel engines and electrical equipment were by no means foolproof. In addition, the high initial cost of these trainsets ensured that they would be used only on a few routes, where passengers willingly paid a premium for speed and luxury. Finally, these trains were ultimately victims of their own success. Because the public was so eager to ride them, demand increased to such an extent that their limited size proved insufficient. Railroads were thus forced to utilize conventional passenger equipment with full-size diesel locomotives in order to maintain comparable schedules.[44]

The benefits of trainsets, especially to EMC, far outweighed these problems, however. While these trains increased the prestige and passenger revenues of the railroads that operated them, they also greatly increased the acceptance of diesel locomotives by American railroads. The favorable public response to these new diesel trains caused railroad managements to consider further diesel locomotive purchases. The presence of even one diesel locomotive required the owning railroad to construct specialized service and repair facilities and train shop forces to maintain diesel locomotives. Once in place, these facilities substantially lowered the effective cost of acquiring additional diesel units, a situation that EMC used to its advantage as the dieselization revolution accelerated.

Diesel-powered trainsets gave EMC a wealth of experience that the company later put to good use in the locomotive industry. This surge of new business, brief though it was, helped convince top GM executives that their tiny (EMC had a staff of no more than two dozen people at this time), largely forgotten subsidiary was neither useless nor moribund. The success of the *Zephyr* and its companions gave Hamilton and Kettering valuable ammunition in their efforts to convince GM to provide funds for the development of a diesel engine specifically for locomotive service.[45]

GM Commits to the Diesel Locomotive Industry

In spite of these successes, Sloan, who preferred to concentrate on lucrative Navy contracts, nearly ended GM's involvement in the locomotive industry. The Model 201 engine had numerous defects, Winton was nearly as recalcitrant as it had been in 1930, and the custom-built nature of the early diesel locomotives plagued a company that had done much to advance the cause of flexible mass production. By persuading Sloan to develop a diesel locomotive engine, Kettering established the key turning point in GM's participation in the locomotive industry. Kettering recalled, "Mr. Sloan called me up one day [in late 1934] and he said 'Ket, we've got to throw this thing out.' "[46] Instead, Kettering asked for a $500,000 research appropriation to develop a new diesel engine for railroad use. He recalled that Sloan's response to this request was: "You know damn well you can't build a locomotive for five hundred thousand dollars," to which Kettering replied: "No, but I think if you get that much in it, you will be likely to finish it." Sloan described this incident in his autobiography, *My Years with General Motors*, but failed to mention that he almost terminated what would eventually become one of GM's most profitable subsidiaries.[47]

Kettering was so persuasive that Sloan soon approved the requested $500,000 for diesel engine research and authorized the construction of a new locomotive production facility. GM gave its final approval in January 1935, and ground-breaking occurred on March 27, 1935. The newly completed facility at La Grange, Illinois (a suburb of Chicago), produced its first locomotive fourteen months later. At first, the facility assembled, rather than manufactured, diesel locomotives, using engines built in Cleveland by Winton and electrical equipment produced by General Electric and other manufacturers. Between 1936 and 1938, EMC equipped its La Grange locomotives with Model 201 engines. By 1938, however, the research program that Sloan had nearly thrown out bore fruit in the form of the far superior Model 567 engine, which replaced the 201 and remained in production until 1966.[48]

GM's decision to enter the diesel locomotive industry was contingent on Kettering's ability to resolve the chronic problems at Winton. During the second half of 1935, Kettering set up a series of monthly meetings to coordinate activities between the GM research labs, Winton, and Electro-Motive. In May 1936, engineers from each of these three groups formed a Diesel Engine Products Study Group to develop engines for various commercial applications, especially for railroad locomotives. At the same time, Kettering increased his control over Winton by transferring its engineering department, including Carl Salisbury and Eugene Kettering, from Cleveland to Detroit. In July 1936, GM's DELCo Products Division began a program to develop new generators and traction motors, based largely on GE designs.[49]

While continuing its research and development on the 567 engine and its related electrical equipment, EMC arranged for the construction of a variety of locomotives, primarily for testing purposes and additionally to demonstrate to the railroads that EMC offered more than just passenger trainsets. EMC's locomotive research and development program consisted of one engineer and two draftsmen, hardly a major financial commitment on the part of its parent company. GE, the St. Louis Car Company, and Bethlehem Steel built the carbodies of the twelve locomotives that were part of this project, while GE and Westinghouse furnished electrical equipment. Electro-Motive completed a two-unit 3,600-hp demonstration model in June 1935, and tested the locomotive on various railroads.[50]

Railroad demand accelerated EMC's initial involvement in the diesel locomotive industry. Although EMC intended its early experimental designs to serve solely as a test bed for locomotive components and subassemblies, two railroads (the Santa Fe and the Baltimore and Ohio) were so anxious to acquire diesels that they demanded to purchase these experimental models. Hamilton relished advanced demand for diesels (as he had with the earlier *Zephyr*), but he was reluctant to sell largely unproven technology. As he commented ruefully, however, "We could not talk them out of this."[51] The Santa Fe units reduced the Chicago to Los Angeles running time of the *Chief* by fifteen hours by operating at speeds of nearly one hundred miles per hour. Despite initial problems—one of the diesels pulling a train carrying Santa Fe president Samuel Bledsoe caught fire on a test run—the Santa Fe was extremely pleased with its new diesels. Given its early experience with diesels and its route through arid and isolated terrain, the Santa Fe soon assumed a leading role in the replacement of steam locomotives with diesels.[52]

Integration and Expansion

Faced with both unexpected railroad demand and increasingly difficult relations with component suppliers, Electro-Motive expanded the La Grange plant and transformed it into a true manufacturing facility. Railroad demand increased as the American economy showed signs of improvement, and EMC output increased from fifty-five units in 1936 to ninety-four units the following year. At the same time, outside suppliers became increasingly unreliable, a situation that threatened to undermine EMC's guarantee policy, as well as its reputation. Hamilton complained that incompetent outside suppliers made it necessary for EMC workers to "get socks and lunch baskets out of the fuel tanks, and socks out of the pipes they had left in there. . . ." He also worried that "even in our own family we got to a point where we weren't getting along with Winton," primarily because Winton was more interested in the development of Navy submarine engines. As a result, in 1936 GM transferred the production of the Model 201 engine from the Winton facilities in Cleveland to EMC at La Grange.[53]

In January 1938, EMC placed the new Model 567 diesel engine in regular production at La Grange. This engine, designed specifically for railroad use, was the end result of the research program that had won Sloan's grudging approval in 1934. It weighed only twenty pounds per horsepower, was far more reliable than the Model 201, and served as the cornerstone of EMC's success for the next quarter century. Although GM designed the Model 567 for locomotive use, GM vice president R. K. Evans, who was in charge of the various diesel engine divisions, believed that it would have "even a greater volume of business in marine and stationary work than in locomotives."[54]

In July 1939, EMC's Transmission Department began building generators and traction motors, and thus no longer relied on GE for these critical components.[55] Since GE had not patented its locomotive electrical equipment technology, EMC simply copied that company's generator and traction motor designs with few modifications. As a result, by the end of the 1930s, EMC had made far greater progress in the integrated manufacture of diesel locomotives than had any of its competitors.

One observer, watching the construction of La Grange, thought that "the production line will resemble that of a modern automobile plant." To a certain extent, this analogy held true.[56] La Grange did have a main "assembly line," the erecting shop, with subassembly lines feeding into it. Because of the enormous size and weight of diesel locomotives, however, EMC modified its "assembly line" by "bringing the tools to the job rather than taking the job to the tools."[57] Small battery-powered trucks trundled back and forth, each carrying a box of tools, a box of electrical parts, and a box of miscellaneous parts. The plant had an on-site testing laboratory to control quality and to help compile data on fuels, lubricants, and new metal-working techniques.

Initially, EMC's La Grange plant manufactured standard-design 600-hp and 1,000-hp yard switchers. The company's decision to introduce yard switchers in 1935 was a sound one. The economy was beginning to improve after the nadir of the Great Depression, yet railroads were still anxious to take advantage of the economies offered by diesel switchers. Their high tractive effort at low speeds made diesels ideally suited for switching service, and, since they seldom strayed far from the yard service, maintenance and emergency repairs could be made with relative ease. Assuming that they were used at least sixteen hours per day, diesel switchers were an excellent investment, usually repaying their initial cost in two to four years.

The use of yard switchers, the first stage of dieselization, required the creation of service and repair facilities at major yards and cities along the railroad, and this in turn facilitated the second and third stages of dieselization, namely, the use of diesels in road passenger and, later, road freight service. Switchers also gave EMC valuable experience regarding diesel locomotive performance in actual operating conditions.[58]

EMC next focused on the road passenger locomotive market. Initially, EMC tailored these locomotives to specific customer requests, producing three similar designs in 1937, all of 1,800 hp. All twelve EA/EB units went to the Baltimore and Ohio, the Santa Fe purchased eleven E-1 units, and the Union Pacific acquired all six E-2 locomotives. Like the earlier 1935 demonstrators, these locomotives employed the Model 201 engine. As EMC became more attuned to the benefits of scale economies (and as demand increased), the company increasingly standardized designs and lengthened production runs. Between November 1938 and February 1940, EMC placed into production the virtually identical E-3, E-4, E-5, and E-6 2,000-hp passenger locomotives, all of which used the new Model 567 engine. Electro-Motive ultimately sold 166 locomotives of these model designations and, after World War II, produced more than eleven-hundred similar E-7, E-8, and E-9 passenger locomotives. More important, these early E-units gave EMC valuable experience in the design, production, and marketing of large road locomotives.[59]

The production of standard-design switchers and passenger locomotives at La Grange prepared the railroads for the dieselization of the largest potential locomotive market, road freight units. Diesel switchers, in particular, impressed railroad executives with their cost savings and these executives, far more than top GM management, urged EMC to build a freight locomotive.[60]

EMC introduced the prototype for a 5,400-hp, four-unit (i.e., four separate locomotives coupled together under unified control) road freight locomotive in November 1939. This locomotive, the first Model FT, was soon in demonstration service on the Santa Fe and, later, the Boston and Maine. Although its four sixteen-cylinder, 1,350-hp engines provided less horsepower than many large steam locomotives, the FT could produce more tractive power than any steam locomotive yet built. The Santa Fe, which purchased its first diesel switcher in February 1935, and its first diesel road passenger unit in August of that year, was so impressed that it soon placed an order for eighty FT locomotives.[61] By March 1941, four railroads, including the Santa Fe, the Southern Railway, and the Milwaukee Road, had placed orders for the FT.[62]

The FT was by no means a perfect locomotive, even though it remained in production until 1945. Richard Dilworth, the locomotive's designer, actually considered the FT to be one of the worst mistakes of his entire career. Mechanical problems occurred with unacceptable frequency. In addition, while the four-unit locomotive could be separated into two two-unit locomotives, each with 2,700 horsepower, one-unit and three-unit combinations were more difficult, since half of the units lacked cabs and shared an electrical system with their cab-equipped companions. This was done partly to save construction costs—units with cabs and locomotive controls were more expensive—and partly to reduce the possibility that railroad unions would demand a complete engine crew on each cab-equipped unit on the train.[63]

While the FT was the most notable achievement of EMC's prewar locomotive development program, the company produced a variety of other locomotive designs, all of them to standard designs. By 1940, in addition to the FT, EMC offered 600-hp and 1,000-hp switchers, a 1,000-hp road switcher, a 2,000-hp transfer unit (which consisted of two 1,000-hp switchers permanently coupled together), and a 2,000-hp road passenger locomotive.

Controlling the Customer

EMC enforced a rigid policy of design standardization, and soon after La Grange opened, the company refused a large order for custom-built locomotives. The company limited the design options of its customers to a very narrow range of choices regarding items such as headlights, horns, fuel tanks, dynamic brakes, and boilers for passenger car heating. This flexibility somewhat resembled the availability of options on passenger automobiles, buses, and trucks, although the correlation between the auto industry and the locomotive industry has been overdrawn by some observers. EMC unquestionably took customer design preferences into account, but only by altering future locomotive designs, not by custom-building locomotives.

Company executives understood that standardization was "necessarily a fundamental policy of this new enterprise if it was to attain the success designed for it."[64] Such standardization, for example, allowed the production costs of diesel switchers to be reduced by approximately 25 percent.

Railroad motive power officials frequently resisted EMC's efforts at locomotive standardization, since it eroded their cherished prerogative of control over locomotive design.[65] EMC resisted their pressure for customized designs in a variety of ways, however. As it had done in the railcar market during the 1920s, EMC sold diesels to high-level railroad executives (many of them trained in finance) rather than to motive power officials. By selling to top officials, EMC could point effectively to the operational cost savings and the high rates of return on investment associated with dieselization. EMC sales forces naturally emphasized that these savings would apply only if railroads agreed to accept standard-design diesel locomotives. EMC used one additional weapon to stymie recalcitrant railroad motive power officials who were determined to retain control over the locomotive design process. On several occasions, EMC designer Richard Dilworth deliberately inserted a glaring error in the diesel locomotive blueprints. When railroad motive power officials met with EMC representatives they quickly spotted the "error." A debate then ensued over the best method of rectifying the situation. Dilworth and other EMC engineers carefully stage-managed the debate to ensure that it would be as acrimonious and lengthy as possible. As Dilworth admitted, "Sometimes the argument lasted 12 or 15 hours, and by that time everybody would be so exhausted that they'd adopt the balance of the layout without further talk."[66]

EMC also found it necessary to persuade its customers of the value of welded subassemblies, especially those used in locomotive frames. Unlike many of its competitors in the locomotive industry, EMC made extensive use of fabricated (welded) components in its locomotives, rather than iron or steel castings.[67] This practice, which was an EMC copy of a Swedish innovation, reduced weight, increased strength, and decreased EMC's dependence on outside suppliers. The use of welded crankcases, for example, reduced weight by about 20 percent. Many railroad motive power officials, relying on traditional practices rather than empirical data, felt that welded frames were ill-suited for locomotive applications. To combat this belief, EMC went so far as to attach structurally unnecessary "dimples" to its welded frames, making them look like cast frames to the casual observer.[68]

During the 1930s, EMC diesel locomotives resembled their ALCo diesel locomotive counterparts in many respects. Both companies offered standardized designs, although EMC was more forceful in its insistence that customers purchase the models that EMC engineers had designed. EMC locomotives performed better than ALCo diesel switchers, but the locomotives offered by both companies were far from perfect. EMC had the more modern production facility but was not yet successful in its attempts to apply jigs, fixtures, and other standardized manufacturing techniques to the production of locomotives. Not until the 1940s would locomotive production become truly efficient at Electro-Motive.

GM Invests in EMC's Marketing Capabilities

EMC's marketing achievements outshone the company's impressive manufacturing advances during the 1930s. EMC's sales staff was never very large—the total number of salesmen in 1948, at the height of the dieselization rush, was only fourteen—but they had a wealth of marketing experience to support them. In its attempt to discover the motive power needs of its customers, EMC was able to take advantage of the resources of GM's Customer Research Department, established in 1933, which comprised the largest market research organization in the world. In order to demonstrate clearly to railroad executives the advantages of diesel locomotives, EMC was the first company in the locomotive industry to develop the "economic study," which provided an extremely accurate estimate of the potential savings of dieselization. This type of study had become standard industry practice by 1947.[69] For railroads that were still undecided, EMC supplied demonstration units free of charge (except for fuel costs), a practice which also became standard throughout the industry. The use of demonstrators was impractical in the steam locomotive industry, since steam locomotives were usually unique.

Unlike the other locomotive builders, EMC realized that the traveling public could play a significant role in a railroad's decision to purchase diesel locomotives. Trains like the *Zephyr* gave the public a taste of what fast, comfortable, soot-free, air-conditioned travel could be like. In addition to displaying diesels at railroad equipment trade shows, EMC featured its products at public venues, such as the 1939 New York World's Fair. The public demand for diesel-powered passenger trains that followed this type of publicity encouraged railroads to purchase diesels for passenger service. These initial sales in turn increased the likelihood that railroads would purchase additional diesels for freight service.[70] Public distaste of smoke, soot, and cinders and concomitant popular support for dieselization indicates that social and cultural values helped influence the pace and direction of diesel locomotive technology.[71]

Once it had sold diesels to a railroad, EMC was able to arrange a number of creative finance options. Just as EMC had leased railcars during the 1920s, the company leased locomotives during the 1930s, stipulating that rental payments be less than the savings produced by the new diesels. In

FIRST-MOVER ADVANTAGES 53

1939, a 600-hp switcher cost $62,250 cash, or $750 per month for eight years, while a 1,000-hp model carried a price tag of $84,300 ($1,000 per month for eight years). Since diesel switchers saved the railroads approximately $1,500 per unit per month, this was essentially a fail-safe proposition. EMC could also arrange financing through the General Motors Acceptance Corporation for railroads unable to secure the bank loans necessary to meet the rental payments or to buy the locomotive outright. This procedure allowed railroads, especially those in receivership, to avoid the equipment trust financing, arranged through banks, that predominated in the sale of steam locomotives. Hamilton had to persuade reluctant GM executives to release some five million dollars in GMAC funds. As he noted, "Their first reaction when I told them what I wanted to do, was very, very, very unfavorable because it didn't make too much sense to talk about selling locomotives to a railroad without any down payment, when it was already in the hands of receivership...." Although GMAC provided the actual financing, EMC managed the loans. In 1937 GMAC financed 12 percent of EMC sales, and this figure jumped to 67 percent the following year.[72] In 1938, when EMC diesel switchers were outselling their competitors ten-to-one, and outselling steam switchers a hundred-to-one, a *Wall Street Journal* article suggested that EMC's success lay in GMAC financing, combined with the ready availability of locomotives from stock.[73]

Initially, railroads could order EMC switchers from stock and expect to receive them in a few weeks, just long enough to apply a coat of the appropriate paint and make a few other minor modifications to suit the needs of a particular railroad. The steam locomotive industry could not match this prompt service. According to Ralph Budd, president of the Burlington, "It is something entirely new for a railroad to be able to order a half dozen switch engines and have them delivered within a week, just as one might purchase highway trucks, but that has been our experience and the experience of other roads since the Electro-Motive plant has been going at La Grange."[74] World War II ended the luxury of building locomotives for stock, and after the war the increased number and variety of locomotive models, combined with the enormous expense of maintaining a locomotive inventory, ended the practice of building for stock once and for all. Nevertheless, the almost instant availability of diesel switchers during the 1930s encouraged many railroads to purchase their first diesel locomotives.

Once EMC had built and delivered a locomotive to a railroad, it offered extensive post-sale support programs. These services, a heritage of Hamilton's involvement in the auto industry, exceeded greatly the limited (or often nonexistent) post-sales services of the steam locomotive industry. Continuing the practice established by Hamilton in the 1920s, EMC provided training for operating and maintenance personnel. Initially, as had been the case in the railcar era, this took the form of traveling instructors who rode with the new locomotive until the regular railroad crews were familiar with its operation.

EMC's instructional services became more formal in 1934, when H. B. Ellis, EMC's service manager, established a locomotive school at the General Motors Institute in Flint, Michigan. Since EMC designed these early classes primarily for the U.S. Navy, the initial response from the railroads was poor; but, as La Grange shipped locomotives to more and more railroads, the popularity of these two-week classes increased. In April 1936, a railroad executive wrote to EMC, suggesting that the locomotive school be moved to La Grange, and EMC did so in 1937. By 1944, the school had trained more than 35,000 railroad employees. In 1937, EMC built a traveling instruction car, housed in a converted passenger car, to eliminate the need for railroad employees to travel to Chicago. This car included a forty-four-seat classroom and a display room that contained cross sections of a diesel engine and other equipment. It could train more than four hundred engineers, firemen, machinists, electricians, or other maintenance employees in one week. Classes at La Grange were thereafter generally reserved for higher-ranking operations and maintenance supervisory personnel only.[75]

EMC added to its training capabilities as diesel locomotive demand increased during the 1940s and 1950s. In 1940, EMC established a Service Training Center, which eventually trained more than two thousand students per year. Between 1941 and 1944, Electro-Motive closed its locomotive school and loaned its instructors to the U.S. Navy. This contribution to the war effort enhanced Electro-Motive's postwar competitive position, since many ex-sailors put their wartime diesel training to use in the railroad industry after V-J Day and were thus already accustomed to Electro-Motive engines. In 1945, Electro-Motive began a "teach the teacher" program, which trained railroad supervisors to create their own instructional courses. Electro-Motive offered its first advanced class (a follow-up to the basic two-week course) in October 1947, and established a sixty-day instructor training class in early 1948. In February 1949, Electro-Motive began a special class for railroad purchases and stores employees and in 1950 added a second traveling instruction car. By 1951, Electro-Motive had trained more than ten thousand railroad employees at La Grange, plus an additional sixty-five thousand in the two instruction cars. Other locomotive builders, particularly ALCo, offered training programs, but the services that EMC pioneered were far superior to those provided by its competitors.[76]

In addition to showing railroads how to operate their new diesels, EMC gave them advice on proper maintenance. Steam locomotive maintenance required highly skilled workers who were trained in the crafts of machining and foundry work and who could custom-fit or even custom-build replacement parts. Diesel locomotive maintenance generally required fewer skilled employees, who replaced parts rather than built them, but did require that

FIRST-MOVER ADVANTAGES

a substantial number of employees be thoroughly trained in internal-combustion and electrical-equipment technology. EMC provided such training through its locomotive school and with traveling instructors. Service manager Ellis also designed an ideal diesel locomotive maintenance shop, the plans of which he furnished to any railroad that requested them. In addition, EMC began a "cooperative educational effort [to foster] a complete change in railroad maintenance practices [and create a] running maintenance program to keep diesels in constant service."[77] The Chairman of the Board of the Louisville and Nashville remarked, "Unlike some locomotive builders who are only interested in the cash sale, the Electro-Motive Corporation furnishes a service which is very valuable and really necessary to secure best results from their power."[78]

The Cost of Market Dominance

Electro-Motive and GM spent a considerable sum in order to attain market dominance and create barriers to entry during the 1930s. GM has never released exact cost figures, but it is possible to piece together a fairly accurate picture of the expenses involved in placing EMC in an effective competitive position in the diesel locomotive industry. In 1930, GM paid $6 million for Winton and an additional $1.2 million for EMC. Between 1928 and 1939, the GM Research Laboratories, with its staff of more than five hundred people, spent approximately $3.3 million on research on diesel engines and related products. Various GM divisions, such as Detroit Diesel and Cleveland Diesel, spent additional money on research and development once Research gave them control over a project. GM Research Laboratories alone spent $408,000 on the development of the Model 201 engine, while Hamilton estimated that GM's entire Model 201 research and development program cost GM approximately $7 million.[79] *Fortune* estimated that GM had spent $25 million on all aspects of two-cycle diesel engine development. The FT diesel freight locomotive cost $500,000 to design, plus another $300,000 to test. The initial La Grange plant cost $6 million in 1935. EMC expanded La Grange almost continuously for the next twenty years, with the added facilities to produce diesel engines and electrical equipment costing about $5 million. Of course, much of GM's diesel engine research and development work was later applicable to the production of marine, stationary, and motor vehicle diesels—a clear illustration of the advantages of economies of scope. Nevertheless, between 1930 and 1940, GM's efforts to enter the diesel locomotive industry cost the company at least $15 million and perhaps as much as $22 million.[80]

This was a relatively small amount of money for the world's largest industrial corporation, even in the midst of the Great Depression. In 1936, when

EMC was beginning switcher production at La Grange, less than 1 percent of GM's total horsepower output was in the form of diesel engines of all types. GM's initial investment was also small compared to the rewards that it might reap. The initial market, to replace all of the steam locomotives in the United States and Canada, was worth more than $4 billion. Moreover, the replacement locomotives would themselves have to be replaced eventually, which promised further revenues. On top of this was a substantial business in spare parts and overhauling and rebuilding services. By the end of 1938 EMC had already sold 225 locomotives, worth $25 million, although the GM subsidiary was still losing money.

EMC did not earn a profit on its operations until 1940, and as late as 1947 Charles E. Wilson, the president of EMC's parent corporation, lamented that La Grange "has as yet not returned one dollar to General Motors that could be used to pay a dividend to a stockholder."[81] GM's willingness to reinvest EMC's profits in the company set it apart from ALCo and Baldwin, two companies that distributed their profits as dividends rather than investing them in new research and development efforts and manufacturing facilities.[82]

Ingredients for EMC's Success

EMC established a first-mover advantage during the 1930s, one that allowed the company to dominate the diesel locomotive industry for the next fifty years, by combining its innovations in manufacturing, marketing, and management with those of General Motors. EMC switcher output increased from 80 units in 1937 to 137 in 1939 and 176 in 1941. By 1936, EMC had attained market dominance in the diesel locomotive industry. Thus, by 1940, even though steam locomotives far outnumbered diesels on American railroads, EMC had created a first-mover advantage that ALCo and Baldwin would never be able to overcome.[83]

Several conclusions are evident from EMC's diesel locomotive research and development program during the 1930s. First, the federal government provided a strong incentive for diesel engine development, in the form of Navy contracts. GM itself admitted that "one big impetus for . . . large multicylinder [diesel] engine development was the Navy contract of November 1932, the railroad application being something of a byproduct."[84]

Second, GM, which became the dominant power in the diesel locomotive industry for nearly half a century, very nearly did not participate in that industry at all. If GM had purchased any diesel engine producer other than Winton, if Winton had not enjoyed a symbiotic relationship with Electro-Motive, if Hamilton had been a less persuasive advocate of diesel locomotives, if Budd had not been so insistent on having a diesel for his *Zephyr*, if passenger-car builders had not serendipitously developed lightweight construction methods, if Kettering had not been so intrigued by diesel technology, if Sloan had been more forceful in his attempts to "throw out" the diesel locomotive program—if *any* one of these events had occurred, the world would likely never have heard of GM diesel locomotives.

Finally, these events show that the process of corporate diversification is neither straightforward nor predictable. GM has rightly been depicted as a leader in the formation of diversified, decentralized companies in the decades following 1920.[85] Because of its leadership role and its status as the world's largest industrial corporation, it is tempting to assume that this diversification proceeded in a planned, orderly manner. For the locomotive industry, at least, this was not the case. Decision-making power flowed through back channels rather than through the official pathways of organizational charts. Wholly owned subsidiaries were often blissfully unconcerned with the needs of their parent company. Customers, like Ralph Budd, did more to push GM into the locomotive industry than did that company's top management. During the early 1930s, at least, historical accident and pure luck played roles as important as careful planning.

IV

ALCo and Baldwin: Established Companies, New Technologies

Two of the three established steam locomotive builders, ALCo and Baldwin, explored the possibilities of diesel locomotives during the 1920s. While GE and Westinghouse entered the diesel locomotive industry to take advantage of economies of scope in the electrical equipment industry, these steam locomotive companies did so in order to maintain the loyalty of their traditional railroad customers. This loyalty was as much a curse as a blessing, however, because it encouraged executives at ALCo and Baldwin to regard the diesel locomotive as adjunct technology, suitable primarily for complementing steam locomotive sales by filling a few specialized niche applications for favored steam locomotive customers. ALCo and Baldwin had little choice in this matter, since each assumed that its established customers might award all of their orders to a competitor if it could not fill their entire motive power requirements. Executives at ALCo and Baldwin did not make adequate investments in new diesel locomotive production facilities and expanded marketing systems during the 1920s and 1930s because they did not believe that diesel locomotive technology would constitute a serious challenge to the steam locomotive industry. Thus, even though ALCo proclaimed itself a pioneer in the diesel locomotive industry, both it and Baldwin lost their potential for market leadership to Electro-Motive, a company that had far more to gain by the eventual dieselization of American railroads.[1]

During the 1920s, executives in the steam locomotive industry had good cause to be suspicious of diesel locomotive technology. Prior to 1930, the size and weight of diesel engines, combined with poor production tolerances, ensured that diesel locomotives were technologically primitive and thus suitable for only a few specialized applications. During the 1930s, however, companies such as GM made impressive advances in diesel engine design and construction. Lightweight, high-strength, specialty alloys lowered the critical weight-to-horsepower ratio, as did the extensive use of welding. Improved manufacturing techniques reduced defects, tightened tolerances, and vastly increased reliability. These improvements, combined with advances in electrical equipment technology, ensured that, by 1940, diesel locomotives possessed enough power and reliability to challenge steam locomotives on all fronts. In addition, the decade of the 1930s witnessed a marketing revolution in the locomotive industry, as Electro-Motive developed standard models, successfully resisted the efforts of railroad motive-power officials to design diesel locomotives, targeted their sales efforts at railroad financial officials, and offered innovative training, financing, and warranty programs.

In several respects, therefore, the 1930s constituted a very narrow window of opportunity in the diesel locomotive industry. Throughout this period, executives at ALCo and Baldwin continued to view the diesel solely as adjunct technology (and continued to do so even well into the 1940s) and therefore did not implement the production and marketing reforms necessary to endow their companies with critical first-mover advantages. As a result, by 1940 these two companies were considerably behind Electro-Motive in the dieselization race. Baldwin, in particular, would find it impossible to catch up.

Brief Market Dominance: ALCo and McIntosh and Seymour

ALCo gained its corporate identity in June 1901, when Pliny Fisk organized The American Locomotive Company for the purpose of acquiring the assets of the Schenectady Locomotive Works and seven other small locomotive builders. These firms were the Rhode Island, Cooke, Brooks, Manchester, Pittsburgh, Richmond, and Dickson Locomotive Works. ALCo later acquired stock in the Locomotive and Machine Company of Montreal (later renamed the Montreal Locomotive Works) in 1904 and in the Rodgers Locomotive Company the following year. ALCo merged with the latter firm in 1909. This consolidation, largely a defensive move against Baldwin's organizational strengths, was part of a great merger wave that occurred at the turn of the century, as firms sought to take advantage of economies of scale and avoid the rigors of "excess" competition.[2] Unlike many of these incipient empires, ALCo was able to rationalize production facilities, eventually concentrating locomotive production in Schenectady, New York, and the company grew and thrived during the early decades of the twentieth century.[3]

ALCo did not enter the diesel locomotive industry until passage of the Kaufman Act in 1923. ALCo produced carbodies for thirty-three diesel locomotives between 1925 and 1931, as part of the GE-IR-ALCo production consortium, but the company had little influence over design or marketing strategies. When GE began production of its own locomotive bodies in 1928, ALCo's role in the original GE-IR-ALCo production consortium ended. In response, ALCo decided to enter the diesel locomotive market on its own. In 1929, ALCo purchased the McIntosh and Seymour Company, based in Auburn, New York, which John E. McIntosh and James A. Seymour had incorporated in 1886 for the purpose of building small stationary steam engines. This firm had begun building diesel engines in 1913, but these were designed for marine and stationary use, applications vastly different from those required by railroad locomotives. In 1931, ALCo delivered the first diesel locomotive equipped with an ALCo-McIntosh and Seymour engine to the Jay Street Connecting Railroad in Brooklyn.[4]

ALCo entered the diesel switcher market at the onset of the Great Depression, and this economic crisis sharply curtailed locomotive sales. During the five years between 1931 and 1935, ALCo sold only eighty-nine steam locomotives and thirty-four diesels. In 1932, ALCo did not produce a single locomotive.[5] The $29 million profit of the 1926–30 period turned into a loss of $13.3 million incurred between 1931 and 1935. The company also lost money in 1938 and 1939.[6]

Still, ALCo's diesels sold well, even during the early years of the Great Depression, until Electro-Motive entered the diesel locomotive market in 1935. In 1934, ALCo had a 73 percent share of the American diesel locomotive market, and a year later its share had risen to 83 percent. In 1936, the first full year of production at Electro-Motive's La Grange facility, ALCo's market share fell to 23 percent. ALCo never fully recovered from this rapid decline of its market share. With the exception of only one year, 1946, ALCo never again attained more than 26 percent of the market, and the company remained a poor second to Electro-Motive throughout its remaining years as a locomotive producer.[7]

In 1940, as ALCo completed the development of a new line of 660-hp and 1,000-hp switchers, the company entered into a new production agreement with GE, under which the two companies would manufacture jointly all diesel locomotives weighing more than one hundred tons. ALCo agreed to use GE electrical equipment exclusively. GE, however, was free to sell its electrical equipment to other diesel locomotive manufacturers. In 1950, the two companies consolidated their sales and service facilities, largely under GE's control. The agreement, which ended in 1953, allowed GE to continue the manufacture of small diesel locomotives and made no provision for a division of the export market.[8]

The 1940 agreement came only a year after EMC began to produce its own electrical equipment, rather than buying GE products. As such, GE saw in its agreement with ALCo a chance to guarantee a market for electrical equipment; while ALCo, well aware of EMC's links to GM, welcomed access to GE's extensive research and development and marketing capabilities. However, by assigning many advertising, sales, training, repair, and servicing responsibilities to GE, ALCo divested itself of the organizational strengths that contributed to success in the diesel locomotive industry. Furthermore, because GE's marketing staff had no innate loyalty to ALCo, they could, and later did, use these marketing capabilities as a wedge to enter the large diesel locomotive market as a direct competitor against ALCo.

Strengths and Weaknesses

The organizational strengths that contributed to ALCo's success in the steam locomotive industry could not be readily adapted to the requirements of diesel locomotive production. More ominously, ALCo's core capabilities in the steam locomotive industry constituted serious weaknesses in the diesel locomotive market.

ALCo's physical facilities were among the burdens of its custom small-batch steam locomotive production heritage. Although ALCo had concentrated all locomotive production in one location in the United States, the Schenectady plant dated well back into the nineteenth century and had always been built, and rebuilt, to suit the requirements of steam locomotive production. After ALCo purchased McIntosh and Seymour, diesel engine production remained at Auburn and was not physically integrated with ALCo's Schenectady operations until the late 1940s.

ALCo's subsidiaries, acquired in the interest of more efficient steam locomotive production, could not be easily transformed into ancillary suppliers of diesel locomotive components. Between 1926 and 1930, ALCo purchased the Railway Steel Spring Company, Heat Transfer Products, and the Jackson Engineering Corporation. ALCo retained a controlling interest in the Montreal Locomotive Works and also maintained a close relationship with the Superheater Company—one of whose founders, Samuel G. Allen, later served as ALCo's president. All of these companies reinforced ALCo's technological and organizational reliance on steam locomotives and on the techniques of small-batch custom production.[9]

ALCo's dividend policies during the 1920s, although perhaps seen as necessary to secure adequate capital for continued steam locomotive production, contributed to the company's financial difficulties during the depression and further eroded its ability to make the investments necessary to exploit the diesel locomotive market. ALCo paid dividends on each share of stock, with a market price of approximately $100, of $6.00 in 1924 and $18.00 in 1925. Between 1926 and 1929, after the post–World War I locomotive boom had ended, ALCo paid $8.00 per share per year in dividends. At the same time, ALCo's net earnings were $7.45 per share in 1926, $4.80 per share in 1927, $1.92 per share in 1928, and $5.40 per share in 1929. Even though ALCo's earnings per share were only $1.41 in 1930, the company still paid a $4.50 dividend. In 1937, a financial analyst deduced that "as a consequence of those disbursements, American Locomotive stock for years was accorded an investment esteem not intrinsically warranted."[10] ALCo thus lacked the financial resources to modernize its plant, establish a diesel locomotive research and development program, and reduce its top-heavy capitalization by buying back its preferred stock.

The Great Depression forced ALCo to suspend dividend payments on the company's 7 percent preferred stock. When payments resumed in 1942, the arrearage had reached $15 million, or $42.75 per share. Nevertheless, ALCo executives could congratulate themselves on avoiding bankruptcy, the fate of their principal competitor in the steam locomotive market, Baldwin Locomotive Works. Although ALCo avoided this financial embarrassment, Baldwin may have come out better in the long run, for its reorganization meant that ALCo's annual preferred dividend obligations, in absolute terms, were nearly ten times those of Baldwin.[11]

Furthermore, ALCo executives found it difficult to transmute their knowledge of steam locomotive technology into a comprehensive understanding of the intricacies of diesel locomotive production. At the same time, ALCo lacked a chairman of the board of directors during the difficult years of the Great Depression. In 1933, William H. Woodin resigned that post to become treasury secretary in the new Roosevelt administration, and the board of directors did not select a new chairman until 1940. This lack of concern for long-term strategic decision-making suggests that ALCo, like Baldwin, virtually ran itself, with production and marketing techniques in the steam locomotive industry being little changed from those used fifty years earlier.[12]

Between 1929 and 1945, during the critical early years of the diesel locomotive industry, ALCo was guided by two chief executives who were well versed in the techniques of customized steam locomotive building but who had less understanding of mass production, design standardization, or internal-combustion technologies. William C. Dickerman served as president of ALCo from 1929 to 1940, when he became chairman of the board of directors. He received a degree in mechanical engineering in 1896 and soon began his career at the Milton Car Works (later American Car and Foundry), where his father was the general manager. During World War I, he directed all ACF production and became vice president in charge of operations in 1919. Throughout his career Dickerman "demonstrated a life-long interest in the technical aspects of the [steam] locomotive, backed by his education as an engineer, in his work in behalf of technical societies, and in many appearances as a lecturer on the subject of railroad motive power."[13]

In 1938, Dickerman thought that "the technical potentialities of the Diesel-electric locomotive are about the same as they were at the beginnings," and that "the possibilities of the Diesel-electric locomotive are already fixed and known . . . [but they are] not so with the steam locomotive." Dickerman went on to list the improvements that would increase the "possibilities" of the steam locomotive. These improvements included roller bearings, integral steel castings, streamlining, superheaters, and coil springs.[14] ALCo executives thus expressed a commitment to technological innovation, but this progress was limited to marginal improvements in traditional steam locomotive technology. Even after ALCo had produced successful diesel switchers, Dickerman believed that diesels would be confined to this specialized application. In an April 1938 address, delivered at a meeting of railroad operating and mechanical officials, ALCo's president explained, "For a century . . . steam has been the principal railroad motive power. It still is and, in my view, will continue to be."[15]

Duncan W. Fraser joined the company in 1901 and served as ALCo's president from 1940 to late 1945 and again from 1950 to 1952. He first held an apprenticeship at the Rhode Island Locomotive Works and transferred to the Montreal Locomotive Works three years later. He rose through the ranks at MLW, eventually becoming works manager, then MLW managing director. He gave up this post in 1920 to become ALCo's vice president in charge of manufacturing.[16] In 1944, Fraser predicted that postwar foreign demand would be almost entirely for steam locomotives. Domestically, he believed that "progress in steam locomotives has gone hand in hand with Diesel developments . . . [and] it is unlikely that there will be any one dominant type of locomotive, at least in the foreseeable future. Steam, Diesel, and electric, each have their advantages." ALCo produced its last steam locomotive four years later.[17]

Executive familiarity with steam locomotive production and marketing techniques helped to create a corporate culture ill-suited for success in the diesel locomotive industry. As was the case at Baldwin, ALCo generally promoted its executives from within the company, with few opportunities for the injection of fresh talent or new attitudes. The training, knowledge, and dedication of ALCo managers undoubtedly contributed to that company's success in the steam locomotive industry. Just as ALCo had difficulty modifying its physical facilities and its other organizational assets, the company only haltingly altered its corporate culture to suit the requirements of diesel locomotive production.

Baldwin's Impressive Lineage

Baldwin traced its earliest origins to 1819, when Matthias W. Baldwin opened a jewelry shop in Philadelphia. The Quakers who constituted a large portion of the city's population apparently believed that such finery was too ostentatious, and the business soon failed. Baldwin then built printing and bookbinding equipment. He constructed his first locomotive (a working model for the Philadelphia Museum) in 1831, followed by a full-size locomotive a year later. By 1837, his company was manufacturing more than forty locomotives per year. Business boomed during the Civil War and grew further as the U.S. railroad network expanded rapidly during the 1870s and 1880s.

Baldwin grew rapidly during the late nineteenth and early twentieth centuries. Matthias Baldwin died in 1866, and the firm changed names at frequent intervals, but it always remained a partnership, rather than a corporation. This did not change until 1909, when the firm, then named Burnham, Williams, & Company, became the Baldwin Locomotive Works, a corporation chartered in the state of Pennsylvania. A need for increased capital motivated this incorporation and not inadequacies in the partnership's organizational structure.[18] Baldwin had produced more than 33,500 steam locomotives during its seventy-eight years as a partnership and by 1910 employed some 19,000 workers and had a capacity to make 2,500 locomotives a year. This capacity was severely tested during World War I, when the company completed an average of seven locomotives per working day. The year 1918 saw 3,580 locomotives leave the Baldwin erecting shops. In spite of the war, Baldwin continued its gradual transfer of operations, begun in 1906, from Philadelphia to the industrial suburb of Eddystone. This process was completed in June 1928, just prior to the onset of the Great Depression and the collapse of the locomotive market.[19]

During the nineteenth and early twentieth centuries Baldwin developed a reputation as a consistently innovative and progressive company. Matthias Baldwin advocated the use of interchangeable parts as early as 1839.[20] In 1842, Baldwin introduced the flexible beam truck, a major advancement in locomotive stability. Baldwin was the first locomotive builder to introduce a two-shift workday and was the first large manufacturing firm in the United States to make extensive use of electrical power in its factory.[21] In 1891, Baldwin's general superintendent, Samuel Vauclain, developed an American version of the four-cylinder compound locomotive, again a major advance in design. The company was the first in the industry to use hydraulic forges. In 1895, Baldwin, in cooperation with Westinghouse, was among the first to develop a mainline electric locomotive. During the late 1800s, Baldwin developed new locomotive designs that increased steadily the size, power, and efficiency of its locomotives and set new standards for steam locomotive design.[22]

Baldwin, more than any other locomotive builder, experimented with nonstandard locomotive products. These offerings brought Baldwin tantalizingly close to developing the technologies that would later be required for diesel locomotive production, albeit a few decades too early. Following the end of the Civil War, Baldwin expanded production of steam locomotives for export, for street railways, and for mine and industrial use. These markets allowed Baldwin to maintain output and employment levels during periods of slack demand and testified to Baldwin's impressive use of flexible manufacturing.[23] However, steam locomotives for street railways (often referred to as "steam dummies") soon proved a technological dead end. Electric streetcars and interurbans offered a more viable alternative, but they required radically different production techniques from those employed in the steam locomotive industry. As a result, Baldwin conceded this market to General Electric and Westinghouse, among others. Those two companies later expanded their knowledge of streetcar and interurban electrical-equipment technology to include diesel locomotive electrical equipment.

While Baldwin eschewed involvement in streetcars and interurbans, the company did develop internal-combustion technology. Baldwin, and other builders, designed steam locomotives for mine and industrial use, but this power source was inappropriate underground or in fire-prone locations. In 1906, through its subsidiary, the Whitcomb Locomotive Company, Baldwin produced what was generally acknowledged as "the first successful gasoline locomotive built in this country."[24] These gasoline-powered locomotives were too small for mainline railroad use, and Baldwin sold most to mines, quarries, and agricultural plantations. Moreover, as mainline steam locomotives became larger and larger, these gasoline-powered locomotives remained small and lightly powered. Because of this, Whitcomb, originally designed to complement Baldwin's mainline steam locomotive production, began to grow technologically, organizationally, and physically (the Whitcomb plant was located in Rochelle, Illinois) distant from Baldwin's regular product line. Whitcomb constructed some 1,230 gasoline-mechanical locomotives between 1910 and 1926, but there was little cross-fertilization between this production and Baldwin's regular steam and diesel locomotive production.[25]

During the 1920s, Baldwin's dividend policies prevented the company from accumulating the cash reserves necessary to finance a comprehensive diesel locomotive research and development program during the 1930s. Baldwin paid out $34.4 million in dividends between 1918 and 1930. During this same period, Baldwin had net earnings of approximately $55 million. Furthermore, the company's board of directors continued to approve large dividends even after steam locomotive orders began to decline in the mid-1920s. There is scant evidence in the financial press of the 1920s to indicate that the steam locomotive industry was moribund, or that Baldwin was either a weak company or an unwise investment. It is unlikely, therefore, that Baldwin was forced to pay these high dividends in order to attract new capital from wary investors.[26]

In its subsidiaries, Baldwin geared investments in production facilities and managerial talent toward the continued production of steam locomotives and found it difficult to transfer these organizational capabilities to diesel locomotive production. The Standard Steel Works Company, Baldwin-owned since 1875, supplied steam locomotive components. In 1926, Baldwin acquired a controlling interest in the Midvale Company, a producer of steam locomotive forgings, ordinance, and structural steel. Baldwin purchased the Southwark Foundry and Machine Company in 1929 and maintained a 32 percent interest in the General Steel Castings Corporation, which supplied Baldwin with cast frames for steam locomotives and their tenders, among other products. When Baldwin began diesel locomotive production, it chose to use General Steel cast frames under these units—after all, this process perpetuated familiar operating routines, railroad officials were familiar with traditional cast frames, and some cost comparisons indicated that cast frames were less expensive to produce than welded ones. Unfortunately, when used in diesel locomotives, cast steel frames tended to crack, a problem that Electro-Motive avoided with its use of welded frames.[27]

Samuel Vauclain

The name most frequently associated with Baldwin was that of Samuel Vauclain. Aside from Matthias Baldwin, Vauclain shaped the destiny of the Baldwin Locomotive Works far more than any other single individual. In 1856 Vauclain was born into the locomotive industry—his father had worked briefly for Matthias Baldwin. He rose through the ranks of the Pennsylvania Railroad at Altoona, where he acquired a reputation as a resourceful and innovative manager. In 1882, when the railroad ordered sixty locomotives from Baldwin, Vauclain went to Philadelphia to supervise their construction. The younger Vauclain joined the Baldwin organization in July of the following year as foreman of Baldwin's 17th Street shops, where he soon reduced the work force by 35 percent yet increased productivity through what he later termed "mass production." This system involved the use of common locomotive components, more efficient scheduling of production, and better employee utilization, yet Baldwin continued to accommodate a bewildering array of steam locomotive designs.[28] Vauclain became a general superintendent in 1886, vice president in 1911, and senior vice president a year later. During World War I, he served as chairman of the Council of National Defense Locomotive and Car Committee and the Special Advisory Committee on Plants and Munitions of the War Industries Board. The board of directors elected Vauclain to the Baldwin presidency in 1919, a post he held until he was replaced by George H. Houston in 1929. At that time, Vauclain became chairman of the board of directors and remained so until his death in February 1940.[29]

Vauclain, who was ultimately responsible for the construction of some 60,000 steam locomotives, was thus firmly schooled in the nineteenth-century techniques of steam locomotive production. In 1926, Vauclain conceded "the established efficiency of modern internal combustion engines," but concluded that "it will be many years before the steam locomotive, owing to its simplicity, its serviceability, and its low production cost, will be relegated to the era of the past."[30] In particular, he believed that a research and development program taking several years and costing millions of dollars would be needed to produce a successful diesel locomotive. A transcript of Vauclain's June 1930 address to the annual convention of the American Railway Association shows that he devoted three pages to steam locomotive development and only one paragraph to diesels. He called the steam locomotive "the greatest of all human devices" and concluded that "we are just beginning to realize what actually can be done with the steam engine . . . that will continue it in service, so that it can be more ably discussed in the year 1980 than at this convention in 1930."[31]

In his long tenure at Baldwin, Vauclain "built up a fine body of engineers, superintendents, foremen, and workmen, to all of whom I sold my ideas. . . ."[32] As Vauclain made clear, other Baldwin executives were trained along similar lines, and most had similar opinions regarding the supremacy of the steam locomotive. In 1932, Baldwin vice president Robert S. Binkerd, in a *Railway Age* article, concluded that steam was the "modern" power suited to virtually all railroad applications. He did concede, however, that gasoline-electrics might be useful in light-duty service and switching.[33] In 1935, the same year that Electro-Motive's La Grange facility opened, *Baldwin Locomotives* contained an article, "Muzzle Not the Ox That Treadeth Out the Corn," that convinced many in the railroad industry that Baldwin executives had deliberately ignored the potential advantages of diesel locomotives. In adapting a biblical passage (Deuteronomy 25:4), Baldwin stressed that steam locomotives were still far superior to diesels and cautioned railroad executives against heedlessly discharging the faithful iron horse that had served them well for the previous century. These remarks, published midway through the critical 1930s, came after Electro-Motive had made impressive advances in diesel locomotive technology, yet before that company had established first-mover dominance.[34]

Vauclain's rigid control over Baldwin and that company's more junior executives created layers of difficulties that would make Baldwin's later entry into the diesel locomotive market far more problematic. First, Vauclain maintained a loyalty not only to the steam locomotive itself, but also to the production and management techniques of the steam locomotive industry. Not until World War II did Baldwin hire or promote executives with training in electrical-equipment or internal-combustion technology. Second, Vauclain's desire for efficient production of ever larger steam locomotives resulted in the construction of the massive Eddystone complex—a poor decision, regardless of the later, unanticipated, decline in steam locomotive orders. Finally, Vauclain, for all of his autocratic power at Baldwin, never challenged the customer-driven practices of the steam locomotive industry. Even after Baldwin commenced diesel locomotive production, the company willingly built diesels to specific customer requests, just as it had done with steam locomotives.[35]

Early Diesel Efforts

Baldwin did take some tentative steps in the direction of diesel locomotive technology, but technological failure and executive attitudes constrained these efforts. In response to the 1923 Kaufman Act, Baldwin built two experimental, and largely unsuccessful, diesel locomotives.[36] Baldwin completed the first of these, construction number 58501, in 1925, at an excessive cost of more than a quarter of a million dollars. Two Baldwin subsidiaries, Baldwin-Southwark and the Midvale Company, supplied the castings and engine components used in the locomotive's construction. Westinghouse supplied the electrical equipment for this 1,000-hp locomotive—appropriate, since the locomotive body was similar to earlier straight-electric locomotives built by Baldwin in cooperation with Westinghouse. The second diesel, number 61000, also of 1,000 hp, emerged from Eddystone in 1929. The German-built Krupp diesel engine installed therein cost Baldwin $100,000. As one motive power expert dourly commented, "work and treasure were poured in without stint," yet expenditures on these two overpriced locomotives did not constitute a comprehensive diesel locomotive research and development program. In 1939, Baldwin finally sold the two locomotives to a New Jersey equipment dealer for a very low price, primarily because the company considered "these two machines a constant reminder of an exploded hope. . . ." Among the various conditions of the sale, Baldwin demanded that its builder's plates be removed from the locomotives, and further, "when they were sold by Baldwin, it was with the understanding that the builder was never to hear of them again. . . ."[37]

The technical problems experienced by these two locomotives undoubtedly dampened any enthusiasm that the company might have had for the diesel locomotive industry, but a more serious problem arose from the reliance of Baldwin executives, particularly Vauclain, on steam locomotive technology and their unwillingness to market diesel locomotives aggressively to railroads. One railway equipment dealer felt that the first two Baldwin diesels were "never pushed by the Baldwin sales force because they were primarily steam power men and looked upon diesel power as a novelty rather than the salvation it has become for the railroads," and that this compounded their high cost and technical inadequacies.[38]

Other evidence supports the dealer's assertion. When the Pennsylvania Railroad requested a demonstration of the 61000 on its lines, Baldwin initially ignored that request. Only after further prodding from the railroad did Vauclain acknowledge that Baldwin did not wish to modify the engine to fit railroad clearance restrictions.[39] It seems unlikely, although possible, that Baldwin designers did not take railroad clearances into account. Vauclain's explanation instead may have been little more than a face-saving excuse to avoid a potentially embarrassing demonstration. Baldwin also ignored an offer by the Cummins Engine Company to supply them with diesel locomotive engines of better quality and at a lower price than those employed in locomotives 58501 and 61000. The president of Cummins wrote, in disgust, "We have heard nothing whatsoever from Baldwin, and we are getting into pretty much of a jam because of this. We have passed up several opportunities to negotiate with the other crowd [the J. G. Brill Company] and have about decided that unless Baldwin shows some sign of life in the near future, we will drop them. . . . It is the first time in our experience that we could not at least get a letter out of people, but it seems impossible to even interest them to this extent."[40] The president of the J. G. Brill Company sympathized, adding that "our friend Vauclain ought to stop fishing down in Florida and get back and get busy with this engine matter."[41]

Baldwin in Bankruptcy

The Great Depression dealt a crippling blow to Baldwin and certainly hampered that company's efforts to participate in the diesel locomotive market. Sales of locomotive products dropped from just over $31 million in 1930 to barely $1 million in 1933. The company as a whole lost money every year from 1931 through 1935. Baldwin stock, which hit an all-time high of $285 per share in 1928, split four-to-one in 1929 at $71.25, but sold for $1.50 a share in 1935. In March 1935, Baldwin defaulted on its bonded debt and declared bankruptcy.[42]

In spite of these financial difficulties, on August 8, 1935, Baldwin filed a reorganization plan that reduced fixed interest payments from $1.3 million to $133,800 per year. Baldwin emerged from bankruptcy court in 1937 and completed its reorganization in September 1938. Baldwin lost money in 1938, but the company earned a profit in 1939, staging what *Business Week* called "one of the most spectacular comebacks in financial history," even though Eddystone still operated at only a fraction of capacity.[43] While Baldwin's insolvency had awakened the company to the need for sweeping changes, these changes occurred too late to ensure Baldwin a significant share of the diesel locomotive market.[44]

Not surprisingly, one of the first developments to occur after Baldwin's bankruptcy was a purge of that company's top management. In March 1938, the company elected a new twelve-member board of directors that retained Samuel Vauclain, largely as a token appointment.[45] The new president of Baldwin, Charles E. Brinley, had little training in diesel locomotive technology. He graduated from Yale, with a degree in mechanical engineering, joined the American Pulley Company in 1901, and became its president in 1919. He remained as president of that company until 1938, when Baldwin decided to make a fresh start with new management. Brinley also served for a time as director of the National Association of Manufacturers.[46]

Baldwin Commits to Diesel Locomotive Production

Following Baldwin's 1935 bankruptcy and even more emphatically after the company's 1938 executive reorganization, managers placed additional emphasis on diesel locomotive production. When depression-induced declines in locomotive orders had encouraged Westinghouse to suspend diesel engine production in 1936, Baldwin considered, and rejected, a Westinghouse offer to sell the production rights to the Beardmore engine. Instead, Baldwin acquired De La Vergne, an older producer of diesel engines but one that had considerably less experience in the locomotive field.

De La Vergne originated as a producer of refrigeration equipment, not of diesel engines. In 1880, John C. De La Vergne established the De La Vergne Refrigerating Machine Company, following successful research into ammonia-based cooling systems. Eleven years later, he acquired a license from an English company, Hornsby-Akroyd, to build an oil engine bearing its name. Following De La Vergne's death in 1896, the company purchased the rights to manufacture gasoline engines from the Koerting Gas Engine Company. That company's founder, a German national named Ernest Koerting, acquired control of De La Vergne in 1909. By 1912, the company had developed a reliable semi-diesel engine, the Model DH. In 1917, the Alien Property Commission seized the German-owned firm and later sold it to William Cramp & Sons Ship & Engine Company. In 1922, De La Vergne introduced its first true diesel engine, the Model VG, designed mainly for pipeline pumping stations. Cramp & Sons left the shipbuilding industry in 1926, forming Cramp-Morris Industrials, an early conglomerate. A year later, it transferred its De La Vergne subsidiary from New York to Philadelphia. Based largely on its proximity to Eddystone, Baldwin purchased De La Vergne and the rest of Cramp-Morris Industrials in 1931.[47]

Because diesel engines did not have any applicability to steam locomotive production, Baldwin did not integrate De La Vergne with its locomotive division. Instead, Baldwin made De La Vergne a part of a subsidiary company, the Baldwin-Southwark Company, which was primarily a builder of hydraulic presses and stamping equipment. De La Vergne engine production was thus organizationally separate from the rest of Baldwin. Baldwin did not fully incorporate De La Vergne into its main corporate structure until 1944; prior to that, De La Vergne sold diesel engines to its parent company at cost. When Baldwin diesel locomotives experienced difficulties in railroad service, the builder sent "a De La Vergne man," rather than a regular Baldwin employee, to remedy the situation. By comparison, Electro-Motive sent its own service technicians to remedy defects in railcars and diesel locomotives, regardless of whether the components had been built by Winton, GE, the St. Louis Car Company, or another outside supplier.[48]

Although De La Vergne had originally designed the Model VO diesel engine for stationary service, Baldwin applied these engines to diesel locomotives beginning in 1935. In late 1936, barely a year after La Grange had begun production, Baldwin completed its first production prototype, a 660-hp locomotive that employed a six-cylinder, four-cycle, Model VO engine and Allis-Chalmers electrical equipment.[49] The VO engine suffered from serious technical limitations, the most severe of which included its excessive size and weight and its high cost of manufacture.[50] The expensive Bosch fuel injectors, purchased from an outside supplier, exhibited serious problems, in several cases malfunctioning on locomotives that had been in service fewer than fifteen days.[51] Baldwin improved the performance of its locomotives during the 1930s, although the company's products were never comparable to those being offered by Electro-Motive.

Beginning in 1939, Baldwin built 660-hp and 900-hp models to a standard design. These locomotives typically used Westinghouse electrical equipment, although the products of other manufacturers could be substituted at the request of the customer—an example of the perpetuation of customer-driven production processes at Baldwin. Baldwin spent some $2 million to put these switchers into production. This sum was insufficient, but it was about all Baldwin could afford given its recent bankruptcy.[52] Altogether, Baldwin sold only twenty-three diesels between 1936 and the end of 1940, and eleven of them were small industrial units.[53]

Even this small output caused Baldwin serious difficulties. Production facilities at Eddystone, while commodious, were not suited to the efficient flow of parts and subassemblies. The new diesels still used the inadequate Model VO engine and continued to rely on crack-prone cast steel frames. In 1940 the Reading Railroad complained that one of its Baldwin locomotives had limped into the shop leaking diesel fuel, cooling water, and lubricating oil and suffered further from a cracked frame, a defective governor, and coolant system failure.[54] These difficulties clearly went beyond ordinary teething problems and did little to inspire customer confidence in Baldwin diesels. Instead, such incidents eroded the precious reservoir of customer loyalty that Baldwin had built up during decades of steam locomotive production.

Many of these design and production difficulties resulted from organizational imperfections at Baldwin. The separation between Baldwin and De La Vergne continued to create difficulties. Additional problems occurred because the Diesel Development Engineering Department had responsibility for all R and D work, while the separate Production Engineering Department dealt with manufacturing. Neither department apparently had much contact with Baldwin's sales force, since, as one Baldwin executive lamented, "the Engineering Department, in the past, has undertaken the development of these engines without any thought of cost or sales prices."[55]

Executive Attitudes and Marketing Strategies

To some extent, Baldwin's new executives understood that Vauclain and his minions had retarded the company's efforts to enter the diesel locomotive market. In 1940, Baldwin executives admitted, "Prior to 1939 The Baldwin Locomotive Works had never made any very active effort to put itself in the position of a builder of competitive Diesel switching locomotives."[56] This admission that Baldwin was extremely tardy in predicting consumer demand for diesels reinforced earlier statements by Baldwin executives who felt that diesels could not compete against steam locomotives on a cost/power basis.

Still, Baldwin executives were not complete converts to the concept of dieselization, in part because they still thought in terms of horsepower rather than operating cost efficiencies. In 1955 a Baldwin vice president, O. DeGray Vanderbilt III, testified "It was obvious *in the early forties* that the diesel was headed for a very important piece of motor [sic] power for the railroads of the country."[57] This statement was part of lengthy testimony given before a Senate subcommittee investigating whether GM and Electro-Motive had conspired to monopolize the locomotive industry. Insulated from Baldwin's stockholders and the general public, and realizing that his company was clinging to life in the locomotive industry, Vanderbilt nevertheless acknowledged that Baldwin managers had not seen the potential of diesels until after Electro-Motive had attained market dominance.

Another problem compounded the hesitancy of Baldwin managers to embrace dieselization. Since Baldwin switchers were inferior to EMC diesels and since Baldwin had higher production costs, Baldwin had to keep its locomotive prices at or below the level set by EMC. This, in turn, ensured that Baldwin lost money on every diesel locomotive that it produced. EMC also lost money on locomotive sales throughout the 1930s, as part of its effort to increase market share. Any strategy to sell at a loss to increase market penetration would likely hinge on the success of an aggressive sales force in soliciting orders and on the ability of the company to sustain considerable short-term losses in one of its major product lines. Baldwin salesmen seemed to have little interest in selling diesel locomotives, however. In August 1939, a railroad executive wrote privately to a colleague that "the representative of the Baldwin [Company], in conversation with me, stated that during the development of the Diesel locomotives they were marketed at a considerable loss, and that his management had expressed the thought that they were in the locomotive business for a profit, and therefore they have not been as aggressive [in their diesel sales] as some of the other manufacturers."[58] This complacency occurred at a particularly unfortunate time in Baldwin's corporate history, since the company was still recovering from its 1935 bankruptcy, and its management was thus extremely reluctant to incur any losses on sales.

This haphazard and cautious approach required Baldwin to shoulder the burden of a typical pioneer, in the form of research and development costs and net losses on sales, yet deprived the company of the rewards that might have resulted from an aggressive and better funded effort to increase market share. As a result, even though it was an early producer of internal combustion locomotives, Baldwin missed the window of opportunity that opened, and closed, during the critical decade of the 1930s.

The Window of Opportunity Closes

The corporate managerial culture and the manufacturing techniques that had made ALCo and Baldwin successful steam locomotive producers prior to 1930 also helped make them unsuccessful diesel locomotive manufacturers afterwards. Their reliance on customer input into the design process, their willingness to offer a multiplicity of designs, and their familiarity with casting and machining practices worked well in the steam locomotive industry. The assumption that what had worked well in the past would work well in the future was central to the managerial strategies of executives throughout the steam locomotive industry. This was not the case, however, since diesel locomotive production demanded very different production and marketing techniques and, above all, very different organizational strategies.[59] By the time these executives realized their mistake, Electro-Motive had created first-mover advantages of such magnitude as to limit seriously the long-term competitive abilities of ALCo and Baldwin.

Given the radically different characteristics of steam and diesel locomotive technology and production, it is hardly surprising that management downplayed the significance of diesels. Steam locomotive orders did decline precipitously during the depression, but this was to be expected in a feast-

or-famine industry. The assumption that a feast of steam locomotive orders was inevitably followed by a famine, followed by another feast, seemed to explain the catastrophic effect of the depression. By adhering to beliefs about the cyclical nature of steam locomotive demand, executives at ALCo and Baldwin essentially denied that technological change during the depression had produced permanent declines in steam locomotive demand. Managers at ALCo and Baldwin misinterpreted the performance capabilities of diesels and erroneously attributed declining steam locomotive orders in the 1930s to a lack of railroad income rather than to a lack of railroad interest.[60]

V

Policy and Production during World War II

WORLD WAR II temporarily stemmed the decline of the established steam locomotive producers but had little long-term impact on the structure of the diesel locomotive industry. As Europe and, later, the United States became embroiled in World War II, American railroads faced unprecedented traffic demands. Western lines in particular struggled to accommodate the massive movements of troops and materiel that accompanied the Pacific campaigns. The Santa Fe was especially anxious to obtain diesel locomotives for use on its arid southwestern routes. The rigorous wartime service that followed removed any lingering doubts in the minds of railroad executives about the superiority of the diesel. Santa Fe executives later concluded, "The war proved the value without a doubt of the diesel locomotive for any type of service. . . ."[1]

World War II greatly stimulated the production of diesel engines for railroad locomotives and for a myriad of other purposes. The diesel engine industry as a whole, which had produced two million horsepower in 1936, turned out fifteen times that amount in 1945. In 1940, 1,111 diesel locomotives were in freight, passenger, and switching service on American railroads, representing one million horsepower. By 1945, 2,864 diesels were in service, with an aggregate of four million horsepower.[2] Diesel locomotive orders for domestic service increased from 160 in 1938 to 937 in 1941. Government production restrictions then caused a decline in orders, to 894 in 1942 and 670 in 1944. Locomotive prices increased as well; a 1,000-hp switcher that cost $79,000 in 1941 cost an additional $10,000 in 1946.[3]

Railroads did not purchase as many diesels as they wanted during World War II, however, because the War Production Board held diesel locomotive production to an artificially low level. Diesels, with their electrical equipment and specialty alloy steel components, required far more strategic materials than did steam locomotives. Despite postwar claims to the contrary, WPB restrictions showed little favoritism to Electro-Motive. Instead, the WPB artificially constrained Electro-Motive's diesel locomotive production, causing the company to lose market share throughout the war. Electro-Motive's enforced stagnation gave ALCo and Baldwin the opportunity, although unrealized, to advance their own R and D and manufacturing techniques. Wartime constraints also gave marginal builders like Baldwin an opportunity to sell far more diesel locomotives than would have been possible under normal conditions. Furthermore, WPB restrictions gave a last breath of life to the dying steam locomotive industry and persuaded some in that industry that the steam locomotive would be reborn in the postwar period. By war's end, even though Electro-Motive had dominated the diesel locomotive industry for a decade, its locomotives had not yet saturated the market, and two-thirds of all U.S. railroads owned no diesels at all—leaving the field open to other producers.

While the actions of the federal government had but a temporary impact on the locomotive industry, the war did produce profound changes in that industry in areas that lay outside the arena of business-government relations. Wartime demand made diesel engine technology stunningly profitable, whether that technology was employed in stationary power plants, ships, or locomotives. This enhanced profitability, combined with orders that exceeded plant capacity, encouraged established producers to expand production, improve manufacturing techniques, and create structured managerial bureaucracies. A leader in this process, the Electro-Motive Company evolved into the Electro-Motive Division (EMD) of GM. Increased wartime demand also persuaded a new producer, Fairbanks-Morse, to enter the diesel locomotive market.

The War Production Board and the Locomotive Industry

Government regulation of the locomotive industry began even before Japan's attack on Pearl Harbor. As early as March 1941, Edward R. Stettinius, director of priorities for the Office of Production Management, gave locomotive builders an A-3 preference rating and placed diesel, gasoline, and electric locomotives on a "priorities critical" list. Stettinius also gave the armed forces priority on all locomotive orders.

In January 1942, President Franklin D. Roosevelt established the War Production Board, under the direction of Sears executive Donald Nelson, to control the allocation of raw materials and wartime production. In April of that year the WPB enacted General Limitation Order L-97 to "control the production and delivery of railroad locomotives."[4] This order assigned production of specific locomotive types to certain builders, based on proven designs. The WPB established a number of locomotive industry advisory committees, the two most important of which were for the producers of large steam locomotives and large diesel locomotives. ALCo, Baldwin, and Lima each sent representatives to the steam locomotive committee, while ALCo, Baldwin, GE, EMD, and Westinghouse participated in the meetings of the diesel locomotive committee. The WPB supplied one representative to each committee from its Transportation Equipment Division. In addition, representatives from several other divisions of the WPB, the War Department, the Office of Defense Transportation, the Office of Price Administration, and other government agencies attended these meetings.[5]

WPB committees established and maintained production schedules for both steam and diesel locomotives. Of the two, the WPB placed more restrictions on diesel locomotive production, since diesels required larger quantities of scarce strategic materials, such as copper and high-strength specialty steels. In addition, the Navy requisitioned a large portion of diesel engine output for use in submarines and patrol boats. The WPB defined five categories of diesel locomotive production: 5,400-hp freight locomotives, 2,000-hp freight locomotives, 1,000-hp road switchers, 1,000-hp yard switchers, and 600-to-660-hp yard switchers. The WPB issued monthly directives for the production of 600-hp and 1,000-hp switchers, as well as quarterly directives for large freight diesels. Electro-Motive produced all authorized 5,400-hp diesels, since it alone had developed designs for these four-unit behemoths. Orders given to one builder could be filled from the available stock of any other builder. Many railroads thus could not obtain diesels from their favored prewar producer (usually Electro-Motive) and many more railroads had no choice but to accept steam locomotives, since so few diesels were available.[6]

All of the major locomotive producers had some degree of influence on government production policies during the war. All companies maintained representatives on either the steam or diesel locomotive industry advisory committees. In addition, both ALCo and Lima had representatives on the WPB Transportation Equipment Division, while a Baldwin executive served with the Defense Transportation Authority.

EMD had no representation at either the Transportation Equipment Division or the Defense Transportation Authority but did have more subtle ways of exerting pressure on the federal government. Ralph Budd, long a staunch supporter of EMD, was an early member of the WPB. He frequently acted as a liaison between the locomotive builders, the railroads, the Office of Production Management and, later, WPB director Donald Nelson. Budd often gave advance advice to railroads concerning the probabilities of successful locomotive orders from various builders. In addition, he was quick to warn EMD, early in the war, that the government might classify diesel locomotives as "nonessential." Budd warned Hamilton that he needed to convince "the people in Washington" of the value of diesels—something that neither Hamilton nor anyone else at EMD had thought to do.[7] Hamilton then encouraged railroad executives to put pressure on the federal government. For example, J. B. Hill, president of the Louisville and Nashville, wrote to Joseph B. Eastman, director of the Office of Defense Transportation, stressing the need for diesels, because "they represent a greater addition to our freight locomotive capacity than we could otherwise obtain."[8] After receiving a copy of Hill's letter, Hamilton replied that "this is exactly the kind of pressure . . . that I had in mind would be effective in Washington."[9]

Budd possessed only a limited ability to help EMD, however. He could not undo WPB regulations, nor could he or any other member of the WPB ensure that EMD would receive the majority of new diesel locomotive orders. Instead, wartime restrictions often worked against EMD. The Louisville and Nashville, which preferred EMD's diesels, was told that these were unavailable, but "It was hinted in Washington, however, last week that if we would be willing to take a 4,000-hp engine of the four-cycle type from the American Locomotive Company, it would be possible to get such an order attended to."[10] WPB officials also harbored suspicions concerning GM's size and financial power. One EMD executive complained that the WPB "did not pay much attention to statements by representatives of the Electro-Motive Corporation [sic] because they feel that their statements are simply prompted by a desire for more business. . . ."[11]

In the minutes of the meetings of the Large Diesel Locomotive Builders Industry Advisory Committee, there is no evidence of favoritism by the WPB toward EMD or any other locomotive producer. The WPB actually forced EMD out of the locomotive industry during the first four months of 1943, since the Navy requisitioned EMD's entire output of diesel engines. Moreover, none of the locomotive producers complained about their treatment under the WPB guidelines. None of the builders thought that General Limitation Order L-97 restricted their development of improved diesel locomotive technology. This acceptance was illustrated by an advisory committee meeting in December 1944, in which the WPB considered ending L-97. Without exception, the representatives of all of the locomotive builders agreed that it should be continued. If any of the locomotive producers had viewed the WPB as unfair or had believed that L-97 prevented them from improving their diesel locomotive technology, they would have undoubtedly called for the revocation of L-97.[12]

Since the WPB placed no restrictions on diesel locomotive research, all of the builders were able to prepare for the postwar period by refining their technological capabilities during wartime. The ability of Fairbanks-Morse to develop a completely diesel locomotive line illustrates during wartime the WPB's willingness to allow extensive R and D activities, even those that would have little positive impact on the war effort. ALCo and Baldwin did not take full advantage of this opportunity, perhaps because each company built more steam locomotives than diesels during the war—ensuring that their management and manufacturing facilities remained focused on the immediate requirements of frenetic steam locomotive production. Their wartime research and development programs were fraught with mistakes and false starts, and these shortcomings did far more than WPB restrictions to hamper the efforts of the two builders to compete against EMD in the postwar era.

When the WPB revoked L-97 in July 1945, the competitive structure of the locomotive industry was identical to the one that existed in 1941. While EMD still dominated the industry, its market share had fallen from 62 percent in 1941 to 47 percent in 1946.[13] Furthermore, market dominance after 1945 was ensured, not by how many diesel locomotives a particular company had produced before that date, but rather by the extent to which a company had established a combined investment in research and development programs, production facilities, and marketing services.

No diesels, not even those produced by EMD, dominated the railroad industry by the end of World War II. In September 15, 1945, the Santa Fe, EMD's best customer, had eighty freight and fifteen diesel passenger units (some 350 total locomotives, usually connected in sets of four), but this was still not extensive saturation in a potential market of some 40,000 locomotives. Between February 1934 and April 1944, railroads placed only 224 diesel passenger locomotives in service—again, just a small fraction of the market potential. By January 1, 1946, only 12.5 percent of the potential road freight diesel market had been filled. Some builders, particularly ALCo, later asserted that EMD still had more of an edge than these figures would suggest, in that railroads tended to standardize their locomotive rosters, so that limited early purchases from EMD would set the pattern for larger later orders.[14]

By war's end, however, nearly two-thirds of all railroads had not yet taken delivery of a single diesel freight locomotive, often because the WPB had forced them to accept technologically obsolete steam locomotives as a wartime emergency measure. In other words, only a third of all U.S. railroads had been given the opportunity to standardize on EMD products before WPB regulation ended. Had it not been for government restrictions, railroads would have purchased even more diesel locomotives than they were actually able to obtain, and most of these orders would have undoubtedly gone to EMD, the dominant prewar producer.

The WPB had much less long-term impact on the American locomotive industry than on other sectors of the U.S. economy, suggesting that the power of government to influence business during wartime may be overrated. John Morton Blum, for example, has suggested that some firms, such as Coca-Cola, Wrigley's, and Henry Kaiser's Six Companies consortium, used the war to their advantage. This is undoubtedly true, but these opportunities did not arise in every industry. Because it was a part of the traditional manufacturing sector, the locomotive industry lacked the rapid growth potential of consumer goods (like soft drinks and chewing gum) and could hardly be considered as important as the high-tech, high-priority synthetic rubber program. As a result, the locomotive industry, like many others, neither benefited nor suffered from its specific interaction with the federal government during World War II.[15]

Electro-Motive: Manufacturing and Managerial Innovations

While government policy did little to alter the competitive structure of the locomotive industry, unprecedented wartime demand for diesel engines had a more profound effect. The success with which the locomotive producers responded to this demand greatly affected their postwar performance. No company responded more quickly, or more comprehensively, than Electro-Motive. For Electro-Motive, World War II was both a blessing and a curse. The war greatly increased its diesel-engine sales, especially for marine use. However, wartime restrictions reduced its share of the locomotive market and gave a temporary advantage to its competitors.

Whereas EMC's success during the 1920s had been primarily the result of marketing strengths, and its success during the 1930s had been largely the result of design standardization, EMD's success during the war years was based on the adoption of standardized manufacturing practices and the integration of the new division into the GM corporate hierarchy. This transition served the company well in the postwar period.

On January 1, 1941, GM reorganized the Electro-Motive Company as the Electromotive Division of General Motors (EMD), largely because GM officials realized the potential for diesel engine sales in a world bent on war. At this time, EMD was directing its efforts more toward marine diesels than to locomotives, and the increased military demand for marine diesel engines led to huge expansion programs at La Grange.[16] For example, between January 1942 and January 1943, EMD enlarged its Model 567 manufacturing facilities by 125 percent. During the war, EMD spent $14 million to add 400,000 square feet of manufacturing space, which could of course be used after the war as well. For the duration, however, EMD was still overburdened and often short of key materials and components. Locomotive production necessarily suffered as a result, and, according to management, "the demands of the Army and Navy for engines and other equipment resulted in the curtailment and eventual discontinuance of locomotives. . . ."[17] Locomotive production stopped entirely during the first four months of 1943. After that, the increased availability of machine tools, combined with the demands of railroad executives for EMD locomotives, convinced the War Production Board to allow locomotive manufacture to resume. In 1944, the WPB increased EMD locomotive production quotas.[18]

Intense wartime demand also placed a premium on production efficiency; here was where EMD used the war to its greatest advantage. Before America's entry into the war, EMC had standardized locomotive designs, but had not yet fully standardized production methods. While Electro-Motive had achieved interchangeability in many key parts and subassemblies (such as those used in diesel engines and electrical equipment), larger components such as underframes were still assembled "piece by piece," with the frequent use of "hand methods." On-the-spot measurements were often more common than the use of jigs and fixtures.[19]

Not until EMD faced unprecedented wartime demand did the division fully adopt mass production techniques. EMD used its WPB-mandated exclusion from the locomotive industry in early 1943 to plan for mass production and to rearrange the facilities and tools at La Grange for the assembly-line production of locomotive components. EMD also reduced the skills needed to assemble locomotive frames and related components by creating a system of jigs and fixtures. These were based on weld-shrinkage statistics that had been compiled at La Grange since the facility opened in 1935, as well as on a detailed examination of the durability of weld joints on wrecked locomotives. EMD developed a just-in-time production and inventory system to allocate the more than 70,000 separate parts that comprised a typical diesel locomotive. Later in the war, EMD constructed separate buildings for the repair and rebuilding of parts and locomotives in order to avoid interference with regular production lines.

Because of EMD's improvements to production facilities and techniques, output increased dramatically. Within six months, freight locomotive production had increased by half, using the same floor space and only a few more employees. By late 1944, EMD was completing two locomotives per day at La Grange, a figure that would more than triple by 1950.[20]

GM understood that its corporate structure should conform to its corporate strategy, and so the company revamped EMD's managerial structure during the war. The creation of the new Electromotive Division in 1941 was the most obvious sign of this transformation. The new division established a line-and-staff organization, with decisions flowing downward through the ranks of factory manager, production superintendents, shift superintendents, general foremen, and shop foremen. The Accounting, Defense Coordination, Engineering, Industrial Relations, Inspection, Material Supply, Sales, and Service Departments performed general staff functions. In addition, new management training programs encouraged employees to "Develop a simple definition of management . . ." and to come up with a list of managerial objectives. By creating an extensive managerial structure, EMD enhanced its ability to respond to changing customer demands and to devise improved production and marketing techniques.[21]

As part of its corporate restructuring, EMD acquired new managerial talent from General Motors. Cyrus R. Osborn transferred from GM's Allison Division in 1943 to become a GM vice president in charge of EMD. Osborn was the first EMD executive to have climbed the traditional GM corporate ladder. Born in 1897, he earned a degree in mechanical engineering from the University of Cincinnati in 1921 and joined DELCo in the same year. In 1925, he became General Manager of the GM Overseas Motors Service Corporation and later served as general manufacturing manager for GM's Export Division. Between 1932 and 1934, Osborn was general manager of GM's Swedish subsidiary, GM Nordiska. In 1937, he became general manager of GM's German subsidiary, Adam Opel, a position, as war approached, with limited potential. In 1941, he became assistant to R. K. Evans, GM vice president in charge of the General Engines Group, which included EMD.[22]

Other GM managers assumed senior positions at EMD. In early 1944, William O. Nelson, who had once been GM's director of purchases in Washington, D.C., transferred from DELCo to EMD to assist in implementing the new production techniques.[23] Nelson C. Dezendorf, who graduated from the University of Oregon and the University of California with degrees in engineering, began his career at GMAC in 1922. In 1931, he became a GMAC vice president and developed a successful new GMAC financing plan two years later. In June 1945, Dezendorf became the new director of sales at EMD. Frank H. Prescott, another DELCo-trained electrical engineer, came to La Grange "to apply mass-production methods to the new motive power."[24]

As the appointment of these executives suggests, EMD reached a managerial turning point during World War II. GM realized that its once-neglected subsidiary had enormous potential and sought to protect and enhance that potential by replacing the manager-engineers at EMC with true corporate men who were schooled not only in the practicalities of internal combustion technology but who also understood that EMD was subservient to the larger goals of GM corporate policy. At the same time, GM suppressed the flexible, experimental, and pioneering culture that had characterized EMC during the 1920s and 1930s. Hamilton had successfully navigated the perilous waters of GM control during the 1930s but, once Electro-Motive became profitable, he lost most of his power—although he continued as a GM vice president. Other holdovers from the early days, such as Richard Dilworth, stayed on, but they had little influence over long-term corporate strategy.[25]

The manufacturing and managerial reorganizations at EMD prepared it well for the postwar dieselization changeover. Once freed of WPB production controls, EMD quickly regained the market share it had lost during the war. Because of wartime demands, EMD had become by far the most efficient producer in the industry and surpassed the efforts of Baldwin, Lima, and Fairbanks-Morse. ALCo was a more formidable competitor, however, partly because of its wartime experience.[26]

ALCo during World War II

ALCo's sales rose dramatically as a result of wartime demand. ALCo production during World War II included more than 2,000 locomotives, 6,000 tanks, 150 boilers for Liberty ships, and many other items. ALCo built 1,354 steam locomotives during 1944, the peak year of steam locomotive production. Employment at Schenectady, which had been 5,500 during World War I, soared to 10,958 in 1943.[27] In spite of ALCo's extensive production of tanks, gun carriages, shells, and bombs, the company kept 60 percent of its locomotive facilities open for railroad use. In 1942 the company produced more diesels than in any prior year. This was followed by another record year in 1943; ALCo's 1944 diesel locomotive tonnage rose 60 percent above the previous year. More important, ALCo's market share increased from 21 percent in 1941 to 40 percent in 1946.[28]

When, in April 1942, the War Production Board allocated production of specific diesel locomotive types, it allowed ALCo to build 2,000-hp freight locomotives, 1,000-hp road switchers, 1,000-hp yard switchers, and 600-to-660-hp yard switchers—four of the WPB's five defined categories. All of these locomotives used the ALCo/McIntosh and Seymour prewar (1940) Model 539 diesel engine.[29]

The WPB placed restrictions only on diesel locomotive production, not on research and development, and during the war ALCo made a concerted effort to design engines of sufficient quality and dependability to compete with those produced by EMD. During the war, ALCo was able to develop a new diesel engine, the 1,500-hp Model 241. The WPB authorized ALCo to install these engines in a set of test locomotives that outwardly resembled the EMD FT. Whereas the four-unit FT developed only 5,400 hp (1,350 hp per unit), a four-unit combination of these ALCo demonstrators would have developed 6,000 hp. In other words, even in the midst of wartime, and during WPB control, ALCo had the opportunity to design and build a prototype road locomotive that could outperform the FT. ALCo wasted this potential advantage, however. The test locomotive (soon dubbed the "Black Maria") debuted in 1945 and was scrapped in 1947—a total failure, for which the Model 241 engine was partly to blame. ALCo only used three diesel engines of this design in railroad service.[30]

Despite this disappointment, ALCo's improved Model 244 diesel engine arose largely from wartime R and D efforts. ALCo introduced the Model 244 engine in 1945, even as the Model-241 equipped "Black Maria" was in demonstration service. In 1948, at the unveiling of an improved version of the Model 244, ALCo President Robert McColl (who had been the company's representative to the WPB) stressed that "the [newly introduced] diesel engine itself has been developed on the principle of constant-pressure turbocharging, the advance of which was greatly accelerated by war developments."[31] And, as ALCo managers explained, "When World War II ended and the railroads flooded manufacturers with orders for diesel-electric locomotives, ALCo was ready with its high-output engine."[32] ALCo emphasized that the Model 244 was technologically superior to any other four-stroke diesel engine. This may have been true, but the Model 244 was less reliable, dirtier, and more difficult to service than EMD's Model 567, which was a two-stroke engine.[33]

In a 1945 report to its stockholders, ALCo claimed that its postwar diesel locomotives would be "a direct outgrowth of wartime research and experience."[34] McColl, the company's president, emphasized that ALCo "can thank war conditions for helping us to develop manufacturing methods which were not usual for us.... By combining our wartime experiences [with those of GE], we have been able to develop a diesel-electric locomotive which, in our opinion, is destined to have an important influence on the diesel-electric locomotive field in the years immediately ahead of us."[35] In 1946, little more than a year after the revocation of L-97, ALCo became the first builder to introduce a successful 1,500-hp road switcher, the RS-2. ALCo thus had a three-year head start in this important postwar market, since EMD did not introduce its first successful road switcher, the GP-7, until 1949.[36]

As these comments suggest, ALCo found little fault with the WPB during and immediately after World War II. By the end of the 1940s, however, when it became clear that ALCo would never rival EMD in market share, its management searched for scapegoats to explain this situation to its disgruntled shareholders. ALCo found one in the WPB. This agency, complained ALCo's management, had shown undue favoritism to EMD and had prevented ALCo from designing or producing a full line of diesel locomotives. In 1949, ALCo executives asserted that "throughout the long period of war, ALCo was prevented ... from making substantial progress in the research and development work which are required for the creation and introduction of an entirely new [diesel engine]."[37] William F. Lewis, an ALCo vice president, reiterated these claims in testimony before a 1955 Senate Subcommittee on Antitrust and Monopoly. Senator Joseph C. O'Mahoney (D-WY), the subcommittee chairman, did not question these assertions, saying that the war "changed the whole direction of the industry—it placed a handicap on the pioneers [like ALCo] and gave an advantage to the newcomers [like EMD]."[38] It is of course questionable to what extent ALCo was a pioneer (i.e., a company that developed first-mover advantages), rather than simply being the first company to build a diesel locomotive. Finally, it should be emphasized that ALCo raised none of these objections during the war itself.

ALCo's increased production during the war afforded many railroads the opportunity to purchase these locomotives and thus compare ALCo's products to their EMD counterparts. By war's end, these comparisons were often in favor of EMD. For example, in May 1941, the Louisville and Nashville saw "little difference" between ALCo, Baldwin, and EMD switchers.[39] In July 1943, the L&N's president preferred the EMD two-cycle engine to the ALCo four-cycle engine but wrote, "I have an open mind on this and our people are giving it a most thorough investigation which will include the operations by some companies who have both types in service."[40] By July 1945, however, the officer reported his company's findings: "The preference of our operating officers as well as the Mechanical Department is decidedly for the Diesel switcher of the Electro-Motive Corporation."[41] During its wartime operations, the Pennsylvania concluded that ALCo diesels exhibited higher operating and maintenance costs than their EMD counterparts. In addition, errors in ALCo production, such as the delivery of two switchers to the Southern Pacific bearing identical road numbers, while minor, reduced customer confidence in ALCo's abilities.[42]

Increased wartime demand also caused ALCo executives to broaden their advertising and public relations campaigns. Prior to the war, the company "never had to give particular attention to what the 'General Public' thinks of ALCo."[43] Because ALCo was involved in large amounts of war work, much of it classified, Board Chairman Dickerman felt it necessary "to reach out through advertising and other means for the attention and respect of a larger part of the general public than the company has needed to cultivate before."[44] However, this advertising merely served as a reminder that ALCo had not yet produced diesel passenger locomotives that could compete against EMD's E- and F-units. An April 1944, ALCo advertisement in a *Railway Age* special issue on streamlined passenger trains showed five such trains, all pulled by ALCo steam locomotives. ALCo emphasized that steam locomotives were perfectly suited to these trains, even though some railroads had begun to dieselize their streamliners ten years earlier.[45]

ALCo began its reconversion process in 1943 and largely completed it by late 1944. The company predicted high levels of both domestic and foreign demand for locomotives and, even during the war, built additional diesel locomotive production facilities. However, since ALCo's executives continued to anticipate a strong postwar demand for steam locomotives, they did not modernize the Schenectady facility to make it better suited for the assembly-line production of diesels. This hesitancy, not WPB restrictions, prevented ALCo from exploiting effectively the first key years of the postwar dieselization boom.[46]

General Electric Hones Its Skills in the Small Diesel Locomotive Market

Like EMD, GE used the intense wartime demand to its advantage by increasing production efficiency.[47] GE did so primarily by forcing railroads to accept standard-design locomotive models, thus cutting delivery time in half. GE noted that railroads learned of the advantages of standardization through this process and soon dropped their insistence on custom-built models. By late 1943, G. W. Wilson, the manager of GE's Transportation Engineering Department, could proclaim that "standardization of locomotives is here to stay."[48] Because of the war, GE began to develop the manufacturing capabilities needed to become a viable producer in the locomotive industry. GE gradually expanded these capabilities after the war, thus establishing the basis for the company's entry into the large mainline freight diesel locomotive market in 1960.[49]

For the time being, however, General Electric manufactured only small switching locomotives for both the domestic and foreign markets under the terms of the 1940 joint-production agreement with ALCo. Several factors made this a sound business decision. First, even though GE executives had no vested interest in preserving steam locomotive production, like their associates at ALCo they believed that diesels lacked rapid growth potential, at least for the foreseeable future. Second, entrance into a diesel locomotive market overcrowded with four producers—EMD, ALCo, Baldwin, and, after 1944, Fairbanks-Morse—entailed considerable risk and seemed to offer few rewards. The cyclical "feast or famine" nature of the locomotive industry compounded this oversaturation problem.

Instead, GE's dual role as a builder of small locomotives and a producer of electrical equipment for larger models entailed few risks. GE could employ economies of scope in order to offset the risks of the locomotive industry by utilizing electrical equipment technology in streetcar, mass transit, and even non-railroad applications.[50] Small diesels—less complicated and less expensive to produce than their mainline counterparts—provided GE with a captive market for electrical equipment and allowed the company to develop and maintain the production facilities, research and development programs, post-sale support services, and customer contacts that could be enlarged at a later date, should dieselization prove more lucrative than originally estimated. By adopting a "wait-and-see" attitude, GE allowed market forces and intracorporate failings to weed out the weaker competitors and, at the same time, demonstrate the magnitude and profitability of the dieselization revolution.

Lost Opportunities at Baldwin

World War II gave Baldwin a rare opportunity to enhance its position in the diesel locomotive industry. Wartime ordnance contracts filled Eddystone to capacity and banished the specter of financial austerity that had accompanied Baldwin's 1935 bankruptcy.[51] When the WPB temporarily ordered EMD to end switcher production, that agency cut Baldwin's competition in this market in half. Since railroads were no longer able to obtain switchers from EMD, they were forced to place Baldwin switchers in service. However, once railroads received their new Baldwin diesel locomotives, they quickly realized that these units were poorly designed and constructed. Far from showing railroads the advantages of Baldwin diesels, wartime deliveries convinced many railroads to avoid further Baldwin orders at all costs.

Baldwin produced a variety of items for the war effort, including stress-testing machines, ship propellers, gun mounts and barrels, tanks and tank destroyers, and both steam and diesel locomotives. Steam locomotive production, especially for military export use, increased dramatically, so much so that Baldwin suspended all ordnance production in early 1944 in order to concentrate further on steam locomotives. In 1944, Baldwin produced 1,200 steam locomotives and only 352 diesels. While these orders temporarily guaranteed Baldwin's financial success, they also ensured that Baldwin's organizational capabilities stayed firmly centered on the production of steam locomotives.

Baldwin expanded its diesel locomotive production capabilities during World War II. In 1944, the company constructed a new diesel engine production facility, and, as a result, diesel engine output during the first four months of 1944 rose 27 percent over the same period a year earlier. Like ALCo, Baldwin was able to develop new diesel locomotive designs during the war, but with even less success. In 1943, Baldwin built an experimental 6,000-hp passenger locomotive that could theoretically outpull the EMD FT. While the FT employed one diesel engine in each of four separate locomotive bodies, Baldwin designed its prototype to accommodate eight engines in a single locomotive shell (only four of these were ever installed). It is not clear why Baldwin tried to pack so much power, and so much equipment, into one locomotive body, but it may have been one more unfortunate legacy of the steam locomotive era, when Baldwin achieved success by designing ever larger and more powerful steam locomotives. In any case, this impractical design did little more than demonstrate Baldwin's inability to produce a viable road locomotive, and it was soon cut up for scrap. Baldwin also collaborated with Westinghouse on the development of a new steam turbine locomotive during the war. This too represented mostly wasted effort, and Baldwin sold only four of these locomotives, the last in 1954. More positively, in January 1945, with Japan's surrender still potentially several years away, Baldwin introduced a new road freight locomotive, complete with two 1,000-hp, eight-cylinder diesel engines and Westinghouse electrical equipment, "in the design of which knowledge gained from war-time operations has been incorporated...."[52]

Unfortunately, Baldwin was its own worst enemy during the war. Railroads, already dubious about accepting diesels not built by ALCo or EMD, found their worst fears justified. Many production defects occurred because of Baldwin's ill-advised attempt to transfer steam locomotive steel-casting technology to the production of diesels. Charles E. Brinley, Baldwin's chairman, admitted to defects in the cast cylinder heads of Baldwin switchers. After repeated attempts to repair a cracked crankcase casting, one railroad maintenance officer wrote in disgust, "The foregoing indicates a very unsatisfactory condition is existing, due to poor material or design, or both, and the builder plainly is responsible."[53] These switchers also suffered from excessive vibrations, and frantic efforts by Baldwin troubleshooters could discover neither cause nor cure. For the Louisville and Nashville, a "kindly feeling" toward Baldwin, its usual steam locomotive supplier, was not enough to offset the evident disadvantages of Baldwin diesels. The railroad's president concluded, "The Baldwin diesel has been less satisfactory in many respects than the other two [ALCo and EMD] locomotives.... Its operating costs are so much higher and along with other unsatisfactory performance that our people are distinctly prejudiced against the purchase of any more Baldwin locomotives at the present time."[54]

Westinghouse owned a block of Baldwin stock and also served as Baldwin's most important outside supplier. As a result, Baldwin frequently found it necessary to negotiate changes in electrical equipment technology and even in the design of its own locomotives. Both Baldwin's Eddystone facility and the Westinghouse East Pittsburgh plant hosted meetings to discuss electrical equipment designs and applications. On occasion, Westinghouse engineers demanded that Baldwin accept and accommodate Westinghouse design changes; for example, the supplier once informed the purchaser that "it is therefore imperative that Baldwin approve E. D. Sk. [Engineering Design Sketch] 149596 immediately...." Westinghouse rebuffed Baldwin's suggestions for electrical equipment redesign, claiming that the cost exceeded what Westinghouse was willing to pay. In frustration, Baldwin approached GE regarding the possibility of substituting some of that company's electrical equipment for Westinghouse products. Baldwin dropped the matter when it discovered that the substitution would require both Baldwin and GE to make extensive design changes. Westinghouse also unofficially marketed Baldwin locomotives by calling on important railroad executives and encouraging them to purchase locomotives manufactured by Baldwin. Finally, even though it acted as an independent supplier to Baldwin, Westinghouse issued its own service and repair manuals for its electrical equipment and provided basic electrical equipment repair in its own facilities.[55]

Westinghouse also shifted managerial talent into the Baldwin organization. In March 1942, Ralph Kelly left his position as vice president in charge of sales at Westinghouse to assume the newly created position of executive vice president at Baldwin. A member of the Harvard class of 1909, Kelly had worked in the Westinghouse Power Engineering Department and the Marine Engineering Division. Between 1934 and 1938, he was a vice president in charge of the operating divisions centered in East Pittsburgh, the plant at which Westinghouse produced locomotive electrical equipment. After serving for a year as executive vice president, Kelly became president of Baldwin in April 1943. He replaced Charles E. Brinley, who resigned the office he had held since 1938 to become chairman of the board, another new position at Baldwin. Other Westinghouse executives accepted positions at Baldwin during and after the war. This injection of managerial expertise produced few benefits for Baldwin, however, since Westinghouse itself had a reputation for being a poorly managed company.[56]

World War II was a lost opportunity for Baldwin. While its share of the diesel locomotive market increased, Baldwin was not able to manufacture efficiently a high-quality product to meet this challenge. Unlike EMD and GE, which streamlined their production and managerial systems to cope with this sudden increase in demand, Baldwin continued to rely on the same production methods that it had employed during the heyday of steam locomotive manufacture. The poor quality of its diesel switchers convinced many railroads to avoid Baldwin locomotives in the future, despite the cordial relationships that they had developed with Baldwin during the days of steam. Although Baldwin's postwar diesel locomotives were much improved over their wartime predecessors, by then the damage had been done.

Fairbanks-Morse: From Railroad Equipment to Railroad Locomotives

Fairbanks-Morse entered the large diesel locomotive industry as a direct result of World War II and the company's attempt to fill excess capacity created by declining war orders. Aside from Lima, it had the shortest tenure in the industry—just fifteen years—and lost money on locomotive sales in all but one of them. A combination of poor management, insufficient resources, an inadequate finished product, and ineffectual marketing programs defeated the company's bid to remain in the diesel locomotive market. In addition, Fairbanks-Morse entered the market too late to be effective. By the time it had begun to climb its learning curve, EMD was already in a position of market dominance.

The origins of Fairbanks-Morse were far removed from the railroad industry. In 1830, Thaddeus Fairbanks invented the platform scale and, together with his brothers Erastus and Joseph, established a factory in St. Johnsbury, Vermont. Charles Hosmer Morse joined the company in 1850 and became a partner in 1866. In 1870, Morse began marketing Eclipse Windmills to a variety of customers, including railroads. Fairbanks-Morse bought the Eclipse Wind Engine Company in 1885, thereby acquiring its plant in Beloit, Wisconsin—a facility that would later be used for the manufacture of diesel locomotives. In 1893, Fairbanks-Morse began marketing gasoline engines and offered small gasoline-powered railroad work cars three years later. The company introduced a two-cycle gasoline marine engine in 1900, followed by a four-cycle version a year later. By 1908, Fairbanks-Morse was building a variety of products, including railroad coaling stations, drawbridge mechanisms, small gasoline-powered railcars and inspection cars, light locomotives, logging cars, lighting systems, track tools, and water pumps and tanks. In the process, the company established a reputation as a supplier of top-quality products for a variety of railroad applications.[57]

Like GE, Fairbanks-Morse explored the possibilities of diesel power early in the twentieth century. In 1912, Fairbanks-Morse established a research laboratory at Beloit to develop a two-cycle, semi-diesel engine; production began a year later. In 1922, Fairbanks-Morse began building large, heavy, marine diesels in both two-cycle and four-cycle variants. In 1931, Col. Robert H. Morse, son of Charles Hosmer Morse, began a research program at Beloit to develop a two-cycle, opposed-piston (OP) engine. As its name suggests, the OP engine had two pistons per cylinder, each at a 180 degree angle from the other. Fairbanks-Morse engineers believed that opposed piston technology offered many advantages, including simpler design, reduced piston speed, easier maintenance access, better heat dissipation, and fewer moving parts. After World War II, Fairbanks-Morse's managers claimed that their company had designed the OP specifically for locomotive use, but this engine was actually better suited for the marine engine market, which was the company's original target. In 1934, the U.S. Navy gave its preliminary approval to the OP engine, with the first installation occurring a year later.[58]

Largely on the strength of OP engine sales, Fairbanks-Morse was the second-largest producer of diesel engines in the United States in 1940. Diesel engines constituted the largest single component of Fairbanks-Morse sales, about a quarter of the total. Although the Navy used OP engines in destroyers, minesweepers, and battleships, the primary application of OPs was in submarines. During World War II, half of the engines that Fairbanks-Morse produced were used in submarines, and half of all submarines built during the war had Fairbanks-Morse engines. The Navy, by far Fairbanks-Morse's best customer, purchased more than 3.5 million horsepower (representing more than three thousand engines) from the company during the war.[59]

In 1939, Fairbanks-Morse used its OP engines in railroad service for the first time. The company installed these 750-hp engines, along with Westinghouse electrical equipment, in six, two-unit railcar sets built by the St. Louis Car Company for the Southern Railway. Whereas EMC had orchestrated railcar production during the 1920s and into the 1930s, Fairbanks-Morse merely sold engines to other railcar producers. Furthermore, early Fairbanks-Morse diesel engines soon proved unsatisfactory in railcar service.[60]

In 1939, shortly after Fairbanks-Morse completed the Southern rail cars, Navy demand increased to such an extent that it precluded the production of OP engines for any other applications. The Navy continued to purchase all available Fairbanks-Morse OP diesel engines until 1944. Wartime demand greatly increased Fairbanks-Morse's production capacity and, between 1941 and 1943, the company spent $5.5 million to expand the Beloit plant for increased production of OP engines. As the company prepared for reconversion, its managers looked for nonmilitary applications for the OP engine. Diesel locomotives were a logical choice.[61]

Other elements of Fairbanks-Morse's wartime experience reinforced the company's interest in the diesel locomotive industry. Fairbanks-Morse opened a diesel engine school in January 1942 to train Navy submarine crews, and this school could also be used to train railroad employees. In addition, the rigors of wartime service gave Fairbanks-Morse engineers valuable experience regarding the OP engine. When Fairbanks-Morse announced its decision to enter the locomotive industry, it asserted that "submarine service has been a most severe test for proving the quality of the new [OP] engine. This and the experience gained in mass production for war needs will have surmounted the two major obstacles facing the new enterprise [of building diesel locomotives]." Presumably, Fairbanks-Morse engineers felt that "the two major obstacles" involved designing an engine capable of withstanding severe operating conditions and then being able to produce it in quantity.[62]

Once Fairbanks-Morse executives had decided to enter the diesel locomotive market, they requested and received permission from the War Production Board to begin research and development work on a new line of diesel locomotives. This was part of Fairbanks-Morse's larger corporate strategy to build both diesel engines and the various types of machinery they powered.⁶³ In May 1944, Fairbanks-Morse publicly announced its plans to build locomotives using OP engines. Although Fairbanks-Morse already built AC and DC electric motors and generators, ranging from one-half to ten thousand horsepower, the company chose to use Westinghouse electrical equipment for its new locomotives.

Fairbanks-Morse selected railroad industry executive John W. Barriger III as the manager of its new Diesel Locomotive Division. Barriger was an ardent advocate of both diesels and "super-railroads," routes that were well maintained and capable of handling large volumes of freight at high speeds.⁶⁴ A 1921 graduate of the Massachusetts Institute of Technology, he had considerable influence in Washington, having served as director of the Railroad Division of the Reconstruction Finance Corporation between 1933 and 1941 and associate director of the Office of Defense Transportation from then until 1943. He was also on the board of directors of the Chicago and Eastern Illinois and Alton railroads.⁶⁵

Although he was an accomplished railroad executive, Barriger could do little to counter the technological and organizational deficiencies that hindered Fairbanks-Morse's ability to succeed in the diesel locomotive market. Fairbanks-Morse's facilities, although large, were not well suited to diesel locomotive production. The company divided production between the new buildings at Beloit and a plant in Freeport, Illinois (the former Stoner Manufacturing Company), which Fairbanks-Morse leased from the federal government in March 1942. Fairbanks-Morse manufactured other components at a plant in Three Rivers, Michigan. The 118-acre Beloit plant was not dedicated solely to the production of locomotives, or even to diesel engines, since it also manufactured pumps, motors, magnetos, lighting systems, water systems, and other items. Beloit's huge foundry, the largest in the world when completed in 1921, was, like similar facilities at Schenectady and Eddystone, not designed for the requirements of diesel locomotive production. Photographs of the diesel-engine assembly area, taken during the war, fail to show anything resembling an assembly line. Finally, although Fairbanks-Morse ultimately planned to offer a full complement of diesel locomotive models, capacity limitations at Beloit restricted actual production to small switchers until well after the war.⁶⁶

In spite of its inadequate facilities, Fairbanks-Morse delivered its first switcher in August 1944. For Fairbanks-Morse, however, the event represented not so much the beginning but rather the beginning of the end. In a market already crowded with three other producers—EMD, ALCo, and Baldwin—even a well-managed company with a sound product would have found it difficult to secure an adequate market share. And, as events during the next fifteen years indicated, Fairbanks-Morse had neither sound management nor a sound product.⁶⁷

The Wartime Experience in Perspective

World War II had a substantial impact on the locomotive industry, not because of government controls, but because the war prompted some builders, especially EMD, to streamline their manufacturing and managerial processes. The War Production Board had little long-term effect on the competitive structure of the American locomotive industry, despite later attempts by ALCo executives to blame that agency for their company's technological and financial difficulties. The government's equanimity did not arise from an indifferent attitude toward the locomotive industry. Efficient wartime transportation required railroads, and railroads demanded locomotives. Instead, the long-established nature of the locomotive industry belied the massive technological changes that occurred in it during the 1930s and 1940s; the industry's location deep within the recesses of the producer-centered capital goods industry made it invisible to most Americans, even though they rode behind the industry's products on a regular basis. As a result, government policy largely ignored the locomotive industry, and policy mavericks chose more glamorous targets. The result was, if not necessarily stagnation, at least minimal transformation.

The altered nature of diesel engine demand had a far more significant effect, however, since the war guaranteed the profitability of diesel engine technology for the first time. GM responded most effectively to this enhanced profit potential, transforming Electro-Motive into a division and implementing manufacturing and managerial controls to speed production. Thus, while EMD's market share declined during the war, thanks to WPB production allocations, the division established a solid foundation for postwar growth. On the other hand, while ALCo temporarily gained market share, courtesy of the WPB, the company did not increase its production efficiency and suffered accordingly after the war's end. Baldwin did far worse, since it translated unprecedented demand into sloppy production and, ultimately, customer alienation. During World War II, Fairbanks-Morse responded both to military demand and to the certain decline of that demand in peacetime. The company's decision to employ its technological expertise and extensive production facilities by entering the diesel locomotive industry in itself constituted a logical business decision, but this decision did not take fully into account the oversaturation of the postwar diesel locomotive market.

After 1945, diesel locomotive demand reached unprecedented levels, exceeding even the growth rates predicted by EMD. Phenomenal growth in diesel locomotive demand accompanied a market shakeout that ultimately proved fatal to most of the diesel locomotive producers. To a large degree, corporate success in the postwar diesel locomotive market depended on the degree to which companies had established technological, manufacturing, managerial, and marketing capabilities during World War II and the decade that preceded it.

VI

Postwar Dieselization and Industry Shakeout

AMERICAN RAILROADS poured millions of dollars into improvements after World War II, and these investments included the purchase of thousands of new diesel locomotives. Substantial wartime profits allowed railroad executives to replace track, bridges, structures, cars, and locomotives that had been worn out by a decade of deferred maintenance during the depression and years of overuse during World War II. Capital improvements to American railroads, which totaled $563 million in 1945, soared to $1.3 billion in 1949. Investments in new locomotives increased from $128 million to $406 million during the same period. The years immediately after World War II thus witnessed a surge of investor confidence in the American railroad industry that belied the later reputation of railroads as inept and conservative.[1]

The advantages of diesels, combined with the increasing price of steam locomotive coal, led to a rapid dieselization of American railroads in the postwar period. Railroads also understood that, in order to achieve the full economies of dieselization, they would have to eliminate *all* of their steam locomotives, along with related service and repair facilities, as quickly as capital constraints would allow. In 1945, diesels handled 8 percent of all passenger train miles, a figure that increased to 22 percent by 1947 and 34 percent a year later. In September 1950, only 17 percent of all freight locomotives were diesels, but these diesels were responsible for 44 percent of freight ton-miles. Domestic steam locomotive orders dropped from eighty-three in 1945 to zero in 1946. In 1953, railroads in the United States owned 14,657 diesels and 15,903 steam locomotives, which although they were in the majority, performed less than a fourth of the work done by diesels.[2] The number of diesel locomotives in service on American railroads increased from 3,882 at the end of 1945 to 20,604 at the end of 1952 to more than 28,000 by the end of 1961. Dieselization was virtually complete by early 1959, with 96.8 percent of all freight ton-miles hauled by diesels.[3]

The fifteen years following the end of World War II witnessed a rapid turnover of firms in the locomotive industry. The dieselization boom temporarily sustained five manufacturers: EMD, ALCo, Baldwin, Fairbanks-Morse, and Lima. Of these, however, only the first two were truly viable producers. By the time Baldwin executives realized that the steam locomotive market had evaporated, it was too late. Lima lacked the financial resources and manufacturing capabilities necessary to compete in the diesel locomotive market, while Fairbanks-Morse executives discovered that the addition of diesel locomotives to their existing product lines stretched their organizational capabilities to the breaking point.

Observers of the locomotive industry understood that not all companies would benefit equally from the postwar dieselization boom. In 1949, an industry analyst predicted that EMD would sell between $900 million and $1.8 billion worth of locomotives over the next decade, with ALCo receiving nearly as much business, $800 million to $1.6 billion. This prediction clearly placed more faith in ALCo than it deserved, for ALCo's subsequent sales were far less than those enjoyed by EMD. The analyst also predicted that Baldwin, Fairbanks-Morse, and Lima would split the remaining $300 to $600 million worth of business.[4]

As it turned out, EMD captured an overwhelming majority of both locomotive sales and the accompanying profits. EMD's market share dipped to a low of 47 percent in 1946 but reached 64 percent in 1950 and 75 percent in 1954. The division's market share reached an all-time high of 89 percent in 1957. Between 1946 and 1959, GM earned an estimated average annual return of 55 percent on its investment in EMD. During the same period, EMD always enjoyed at least an 11 percent return on sales, with an average of 20 percent. ALCo, however, lost money in four of these years and averaged only a 2 percent annual return on sales. EMD earned a 144 percent return on investment during the 1950s, with ALCo earning 9 percent and Fairbanks-Morse losing 14 percent during this decade. In 1951, one of the best years in the entire history of the locomotive industry, EMD earned a 269 percent return on its investment in plant and equipment. In the same year, ALCo managed only a 28 percent rate of return, while Baldwin and Fairbanks-Morse lost money.[5] In retrospect, EMD's success is hardly surprising, since the division was the only fully integrated diesel locomotive producer, its executives had no commitment to steam locomotive technology, all of its production facilities had been designed specifically for the manufacture of diesel locomotives, and it was the only builder that then offered reliable high-horsepower road freight and passenger locomotives.

Embracing Dieselization: The Pennsylvania Railroad as a Case Study

For the far-flung Pennsylvania Railroad (PRR), the decision to dieselize was not an easy one. While its main line extended from Philadelphia to Chicago via Pittsburgh, the railroad also served such major cities as New York, St. Louis, and Washington, D.C. Since coal was by far the PRR's largest source of traffic, and since the railroad maintained investments in numerous coal mines, company executives wished to retain coal-fired steam power as long as possible. Motive power officials understood steam locomotive maintenance and construction practices (the railroad built its own steam locomotives at its massive Altoona, Pennsylvania, shops) and they believed that more powerful steam locomotives could hold diesels at bay. Soaring operating expenses and the inability of PRR motive power experts to design truly efficient steam locomotives caused the Pennsylvania to question its loyalty to steam; but it was not until a younger, more dispassionate group of managers arrived on the scene that the company finally accepted dieselization.[6]

Since the PRR served many densely populated areas, senior executives responded to public and legislative pressure for cleaner motive power. One trackside resident pointed out that seventy tons of soot fell on her town (Tyrone, Pennsylvania) each month, adding that "I always shudder when putting out a wash when I here [sic] one of the numerous freightsplugging [sic] by . . . ," and suggested that the PRR purchase diesels as soon as possible.[7] PRR executives expressed concern that "prosecution has been threatened for violation of the Sanitary Code" as a result of continued steam locomotive operations on Long Island and concluded that "the substitution of these diesels will improve public relations in this territory and eliminate the complaints continuously being made about smoke."[8]

PRR officials sympathized with the attitudes of its passengers, an understandable situation given the precipitous decline in the demand for long-distance passenger travel following the end of World War II. At a meeting of the PRR board of directors, John F. Deasy, vice president-operations, pointed out that "the diesel engines have been advertised—they have glamour and allure to the public. . . ."[9] Martin Clement, the railroad's president, stated that, with regard to passenger traffic, "the thinking in the Diesel situation seems to be that certainly we should be protected in our principal passenger trains so that none of our competitors can have a situation better or equal to that which we have in through passenger service, and they won't have when we have the Diesels."[10] In stagnant or even declining postwar freight and passenger markets, the PRR simply could not afford to ignore the technological advances offered by dieselization. Many of the PRR's freight customers also pressured the railroad to dieselize as soon as possible. In the Philadelphia area, Texaco temporarily suspended the transfer of petroleum products while steam locomotives switched cars at the facility, and Sun Oil refused to allow steam locomotives onto its property.[11] The Ford Motor Company plant at Metuchen, New Jersey, forbade the use of steam locomotives, which would have triggered automatic sprinkler systems. All of these companies requested that the PRR purchase diesels as soon as possible.[12]

The inefficiency and expense associated with continued steam locomotive operations provided the strongest rationale for rapid dieselization. Various PRR studies indicated that the annual cost savings associated with dieselization ranged from 20–40 percent of the cost of the initial investment. In one case, the railroad estimated annual savings of 52 percent—enough to repay the cost of diesels in less than two years. Most of these savings accrued from the replacement of inefficient steam locomotives with smaller numbers of more reliable, although not necessarily more powerful, diesels.[13] Additional savings resulted from the closure of steam locomotive servicing and repair facilities, although PRR officials understood that these savings occurred only when diesels had completely replaced steam locomotives in a particular area.[14]

The Pennsylvania carefully evaluated the performance of various locomotive models on other railroads, as well as on the Pennsylvania itself. PRR test data suggested that total operating and maintenance costs for a typical Electro-Motive unit amounted to $2.32 per hour, while comparable figures for ALCo and Baldwin averaged $2.40 and $2.42, respectively.[15] Input from other railroads reinforced the PRR's favorable impression of Electro-Motive's lower operating costs. The Santa Fe, an early purchaser of Electro-Motive freight diesels, forwarded test data to the Pennsylvania and strongly endorsed the merits of Electro-Motive diesels.[16] The PRR concluded that "the Diesel engines offered by Electro-Motive . . . have greater capacity for the work required than does either of the engines offered by American or Baldwin."[17] Statistical data supported these conclusions. By January 1948, nearly 60 percent of locomotive-related passenger train detentions were attributable to Baldwin diesels, compared to 27 percent for EMD units, even though the PRR operated more EMD locomotives than Baldwin diesels in passenger service. Baldwin diesels experienced a mechanical or electrical failure every 6,069 miles, on average, while EMD locomotives averaged 77,610 miles without incident. Some of the worst offenders were the Baldwin DR-12-8-1500/2 "Centipede" diesels (so called because of their twenty-four wheels), which gave very poor service. Not surprisingly, the PRR concluded that Baldwin diesels were extremely unreliable.[18]

By the late 1950s, when the PRR was phasing out its last steam locomotives, the maintenance costs of EMD-built freight units ranked well below those of its competitors. During the first eight months of 1957, the PRR spent twenty-two cents per locomotive mile to maintain EMD units, while spending more than three times as much—seventy-three cents per locomotive mile—on ALCo units. Comparable figures for Baldwin and Fairbanks-Morse diesels equaled sixty-five cents and fifty-eight cents, respectively.[19] The PRR scheduled preventative maintenance overhauls for its EMD units every thirty-four months, compared to every eighteen months for ALCo diesels, and every twenty-seven months for Baldwin and Fairbanks-Morse diesels. By 1958, PRR executives had concluded that "while the late model Alco units are giving satisfactory service, and their maintenance costs cannot be considered excessive, nevertheless they are still not comparable with their General Motors competition."[20]

Two significant factors characterize the Pennsylvania's postwar diesel locomotive purchases. First, since the PRR and Baldwin had a long tradition of cooperation in steam and electric locomotive production, the former company was willing to subsidize the latter to some extent. As such, despite EMD's superior quality, the PRR felt "that this should revert to a question of policy as to distribution [of locomotive orders] that would be in the best interests of the Pennsylvania Railroad . . . it might be proper to place 5 locomotives with Electro-Motive, 5 with American, and 9 with Baldwin. . . ."[21] Still, loyalty had its limits, and by late 1948 the PRR was ordering nearly 40 percent of its total diesel locomotive horsepower from EMD, 35 percent from Baldwin, 16 percent from Fairbanks-Morse, and barely 7 percent from ALCo.

Second, PRR executives were willing to distribute orders among a variety of builders in order to prevent EMD from monopolizing the locomotive industry, hoping that Baldwin and ALCo would eventually offer better-quality locomotives, or at least locomotives that better suited the railroad's needs. Neither ALCo nor Baldwin did so during the late 1940s or early 1950s, however, and PRR managers concluded that continued profitability and the long-term survival of their company demanded that they invest their limited capital where it would earn the highest possible rate of return—that is, by purchasing EMD locomotives almost exclusively.[22]

Stability Replaces Experimentation at EMD

The years between 1945 and 1957 were golden ones for EMD. Investments in manufacturing facilities and the creation of a marketing and managerial system during the Great Depression and World War II began to pay substantial dividends. Employment at EMD, which had been only four hundred in 1936, increased to twelve thousand by 1947, while the annual payroll rose from $117,000 to $43.5 million in the same period. By 1947, La Grange represented an $80 million capital investment, including inventory, with additional millions invested in EMD's service and parts distribution network. La Grange had produced more than $400 million worth of locomotives by 1947, and orders continued to flood the factory. This output, combined with production at ancillary plants in Chicago and Cleveland, allowed EMD to produce two hundred locomotives per month by 1950. In March 1954, less than eighteen years after La Grange began production, EMD completed its fifteen-thousandth diesel locomotive.[23] Between 1946 and 1959, EMD built 17,343 locomotives, valued at $2.7 billion.[24]

Much of EMD's success in the postwar period resulted from the popularity of its new locomotive models. In 1945, EMD began production of the 1,500-hp F-3, an improved version of the FT. The even more popular F-7 followed in 1949, and both models gained widespread use in freight and passenger service. The F-3 and F-7 were rugged, versatile, and reliable.

EMD's greatest success came in the emerging road switcher market, however. ALCo and Baldwin had a head start in this area, since each company introduced a 1,500-hp road switcher in 1946, barely a year after WPB controls were lifted. Furthermore, EMD's first road switcher offering, the 1,500-hp BL-2, proved disappointing, and only fifty-nine were sold. In November 1949, three years after ALCo and Baldwin introduced their road switchers, EMD introduced the GP-7, a 1,500-hp locomotive that was powerful enough for mainline service, light enough for low-density branch lines, and provided engine crews with excellent visibility. The nearly identical GP-9, introduced in 1953, offered increased horsepower. U.S. and Canadian railroads ordered nearly seven thousand GP-7 and GP-9 locomotives, erasing the three-year lead of ALCo and Baldwin almost overnight.[25]

Although EMD increased the size of La Grange without interruption between 1935 and 1958, the division refrained from expanding its facilities to such an extent that it could accommodate all of the locomotive orders from the entire railroad industry. Had it done so, EMD could have driven all of its competitors from the industry, creating two problems. First, the federal government was likely to take a dim view of any company possessing a 100 percent market share and did eventually prosecute EMD for alleged antitrust violations during the 1960s. Second, when the dieselization boom began its inevitable downturn, EMD could retrench easily and allow its smaller, high-cost competitors to suffer the brunt of declining demand.[26]

In order to relieve the production overload at La Grange and avoid overexpansion of that facility, EMD opened two ancillary locomotive plants. EMD acquired Plant #2, located in South Chicago, in May 1946, in order to increase its production of locomotive subassemblies. In 1951 EMD installed its first conveyor-belt assembly line in Plant #2 to apply rust-proofing materials to locomotive cabs. This use of conveyor belts was the exception, however, and EMD generally used overhead cranes to transport large locomotive components. In January 1949, EMD began production of 600-hp and 1,000-hp switchers, using 60 percent of a 460,000-square-foot facility that the U.S. Navy had built in 1942 to increase diesel marine engine production at GM's Cleveland Diesel Engine Division. By 1950, three thousand people worked at Plant #2, with half that number at Cleveland, in addition to ten thousand employees at La Grange.[27]

Along with these improvements in manufacturing, EMD expanded its marketing capabilities during the postwar period and continued to recognize the importance of public opinion in the locomotive industry. The division targeted its 1946 "Better Trains Follow Better Locomotives" and

"Bright New Era" ad campaigns at railroad passengers. In May 1947, EMD dedicated the GM *Train of Tomorrow* by launching it on a six-month tour of more than thirty cities in the United States. GM never intended this combination of a single diesel locomotive and four passenger cars to be a regular production item at EMD. Instead, GM President Charles E. Wilson was desirous of "creating a greater interest in rail transportation by the public and a greater acceptance of the products we furnish to railroads."[28] A few years later, EMD issued a thirty-two-page booklet, targeted at the general public, that stressed the virtues of diesel locomotives in general and EMD locomotives in particular. EMD executives thus understood, as they had since the 1930s, that public pressure could be a significant factor in encouraging railroads to replace their dirty and noisy steam locomotives with modern diesels.[29]

EMD also played on the vanity of senior railroad executives. The division supplied them with handsomely framed pictures of EMD diesels for their office walls. More important, its highly successful 1957–59 advertising campaign, "Men Who Build the Future of American Railroads," placed color portraits of important railroad executives on the front cover of *Railway Age*, the leading industry trade journal. EMD locomotives also appeared on these covers but in the background.[30] While early EMD advertisements had stressed "super-production and super-transportation," by the late 1940s EMD had shifted its advertising emphasis from "selling the concept of the diesel-electric to selling complete and total dieselization." Shortly after the war, EMD developed the Operating Comparison Report to serve as a clearinghouse for information from more than thirty railroads regarding the most efficient utilization of EMD diesel locomotives. Some railroad motive-power officials thought that these reports were of little value, since most railroads had vastly different operating characteristics. Nevertheless, these services often impressed the senior railroad executives who made the final locomotive purchasing decisions.[31]

After the war, EMD's Service Department expanded its training programs. To back up its approximately twenty salesmen in the United States, EMD employed 120 locomotive instructors, a figure that reflected the value of these instructional courses to continued locomotive sales. In addition to its two instruction cars and its classes at La Grange, EMD sent a complete diesel locomotive on an instructional tour of American railroads. In 1953, recognizing the increasing importance of foreign orders, EMD instituted a ninety-day export training class. Gradually, as larger numbers of railroad operating and maintenance employees became familiar with diesel locomotives, the importance of these instructional services declined. For many years, however, EMD's locomotive school increased goodwill and accustomed a generation of railroad employees to EMD locomotives.[32]

EMD expanded its parts and service network to accommodate the growing number of diesel locomotives on American railroads. In February 1949, EMD opened a new parts distribution center at La Grange, containing some $3 million in inventory. EMD also operated parts distribution centers in other cities, including Jacksonville; Halethorpe (Baltimore), Maryland; Minneapolis; Los Angeles; Emeryville, California; and St. Louis. By the end of 1949, EMD had invested $39 million in its parts distribution network, an amount that was nearly double what ALCo had expended on its entire postwar plant modernization program.[33]

EMD also established locomotive rebuilding locations throughout the United States. Division managers believed that locomotive rebuilding would increase goodwill, provide a captive market for EMD parts, allow greater amortization of research and development and tool and die costs, and serve as a test bed for experimental techniques that, if successful, could be implemented at La Grange. A former dog biscuit factory in Oakland, California, housed the first rebuilding facility, established in 1944. Facilities in Jacksonville, Baltimore, and Emeryville, California, soon followed.[34]

Executives trained in sales, rather than production techniques, became increasingly dominant as older manager-engineers retired from EMD. In early 1950, Richard Dilworth retired from EMD, although he continued to serve as a consultant until 1952. Dilworth was among the last of the informally trained manager-engineers left at EMD. The year 1953 saw the departure of O. F. Brookmeyer, another manager of this type.[35] Later EMD executives, such as B. A. Dollens and Richard Terell, both of whom served as GM vice presidents in charge of EMD, were lifelong GM employees who lacked a particular commitment to diesel locomotive technology.[36]

Between 1945 and 1957, Electro-Motive reaped the benefits of its 1920s decision to stress the development of marketing capabilities, of its 1930s investments in research and development, and of its wartime efforts to refine its manufacturing and managerial techniques. EMD did more than rest on its laurels, however. The division expanded its successful marketing programs and, more important, completed the transition from a corporate culture based on experimentation to one based on stability and predictable profitability. This endeavor had its limits; after all, the locomotive market constituted one of the least stable sectors of an inherently unstable capital goods industry. Nevertheless, by exploiting economies of scope, EMD shifted resources away from locomotive production and toward the manufacture of stationary and marine diesels as railroad demand decreased. Similarly, EMD's parent company, GM, shifted its resources away from EMD and toward the production of diesel-powered busses, trucks, and construction equipment. As a result, fluctuations in diesel locomotive demand had relatively little effect on either EMD or GM.

ALCo Tries to Regain the Lead

ALCo executives could be cautiously optimistic in 1945, with their company poised on the threshold between frenetic wartime production and the postwar dieselization boom. War-related research and development had enabled ALCo to improve the performance of its diesel engines and diesel locomotives. ALCo's Schenectady plant, while neither as modern nor as well planned as La Grange, was nonetheless far superior to Baldwin's white elephant at Eddystone. Finally, ALCo possessed considerable financial resources, which included substantial cash reserves and valuable investments in subsidiary companies. For all these reasons, ALCo in 1945 appeared capable of mounting a sustained challenge to EMD's market dominance or at least of continuing indefinitely as a viable competitor in the diesel locomotive industry.

Appearances proved deceiving, however. After 1945, ALCo began a slow decline that, by 1960, had become a virtual free fall that culminated in the company's exit from the diesel locomotive industry in 1969. ALCo waited too long before updating its manufacturing and marketing techniques to correspond with advances in diesel locomotive technology. ALCo placed greater reliance on its production partner, General Electric, for critical locomotive components and marketing expertise, thus weakening ALCo's organizational capabilities. ALCo proved less adept at managing economies of scope, less able to move resources in and out of diesel locomotive production as market conditions varied. Postwar diversification did little to create economies of scale. Instead, diversification encouraged ALCo to become even more committed to customized, small-batch, production. Custom production had served ALCo well as a steam locomotive producer, but the practice bred catastrophic failure in the postwar capital equipment market. The confusion and discontinuity created when nineteen officers and directors of the company retired, resigned, or died between 1947 and 1954, the years of peak diesel locomotive demand, exacerbated these problems.[37]

Management and Labor

ALCo began the postwar period with a managerial reorganization. In December 1945, Duncan Fraser resigned his position as president and assumed new responsibilities as board chairman. Executive Vice President Robert McColl took Fraser's place as president. At the same time, William Dickerman resigned his position as chairman of the board (in order to make way for Fraser), although he continued as a regular board member. Dickerman's demotion probably had more to do with ill health (he died less than a year later) than with his lack of familiarity with diesel locomotive technology. Still, McColl, as president, was more willing to embrace diesel locomotive technology. In January 1947, little more than a year after assuming his new duties, McColl predicted that 93 percent of all domestic locomotive orders would be for diesels.[38]

In spite of ALCo's new appreciation for diesel locomotive technology, the company continued to allocate resources to steam locomotive production—a further example of the constraints imposed on established firms as they struggle to respond to radical technological change. In August 1945, a memorandum issued to shareholders predicted "that great advances will be made in the development of coal-burning steam locomotives during the next decade."[39] Four months later, ALCo Senior Vice President Joseph B. Ennis asserted that "the future holds an expanding role for both the steam locomotive and the Diesel."[40] ALCo's advertising policy reflected these attitudes. In October 1945, ALCo promoted its steam locomotives with ads bearing the heading "Alco: The Mark of Modern Locomotion."[41] In 1946, ALCo's weekly advertisements in *Railway Age* still alternated between portrayals of steam and diesel locomotives, even though U.S. railroads ordered fifty-five steam locomotives that year (thirty-six of them from ALCo), compared to nearly nine hundred diesels.[42]

ALCo's commitment to continued steam locomotive production continued into the late 1940s. As late as December 1947, ALCo's Junior Engineer Training Program instructed trainees in the techniques of steam locomotive construction. In the same year, ALCo exhibited an all-welded steam locomotive boiler—the latest in steam locomotive technology—at the Railway Supply Manufacturers Association convention in Atlantic City. W. A. Callison, an ALCo vice president, stated, in 1947, "We do not, by any means, believe that the steam locomotive is dead."[43] In 1948, as ALCo turned out its last steam locomotive, Perry T. Egbert, then vice president in charge of the locomotive division and later president of the company, announced that "American Locomotive is not intentionally going out of the steam locomotive business. It is simply a matter of [lack of] demand."[44]

ALCo did not gain a chief executive with significant experience with diesel locomotives until December 1952, when Perry Egbert became president of the company. He received a mechanical engineering degree from Cornell University in 1915 and later served in the U.S. Army Air Corps. He taught experimental engineering at Cornell in 1919 and 1920 and in 1921 became ALCo's technical representative in East Asia. In 1929, following the acquisition of McIntosh and Seymour, he took control of ALCo's diesel-engine development program. Egbert became manager of railroad diesel sales in 1934 and vice president in charge of diesel locomotive sales ten years later. While there is no explicit evidence to suggest that senior ALCo executives punished Egbert for his loyalty to diesel locomotive technology during this time, it is nonetheless suggestive that he spent ten years (1934–44) in the same position; even in the latter year he received what amounted to little more than a horizontal promotion. Egbert eventually directed ALCo's postwar conversion from steam to diesel locomotive production and became president when ALCo's directors realized that market conditions had imposed a product transition that Egbert had advocated for decades. ALCo now had a president who understood how to exploit the dieselization boom; yet, even as Egbert began his new duties, that boom was ending.[45]

Labor disputes and high wage rates also put ALCo at a serious competitive disadvantage.[46] Between 1943 and 1960, at least eighteen separate strikes or work slowdowns interrupted ALCo's locomotive production. These ranged from a two-day strike by two hundred workers in 1943, to a two-month strike in 1946, to a three-month strike in 1951, to a six-month strike in 1952–53. Concessions made by ALCo executives to the United Steelworkers of America contributed to the company's high wage rates, as did overuse of piecework incentives. In December 1951, average wage rates at GM were $1.91 per hour. ALCo's workers at Schenectady earned $2.41 per hour. To produce a diesel locomotive, ALCo spent fifty-eight cents more in wages, per hour, than the industry average.[47]

To some degree, these labor difficulties were beyond the management's control—the USWA was, in general, a more contentious union than the UAW, which represented EMD's workers. More broadly, worker dissent arose from ALCo's unavoidable devaluation of traditional skills in casting and machining in favor of those in welding, diesel mechanics, hydraulics, and electrical work.

ALCo's Postwar Strategy

As the war neared its end, ALCo's ordnance production virtually ceased and seemed unlikely to reappear. ALCo executives, daunted at the prospect of wartime facilities standing empty and unused, adopted a dual strategy of expansion.

First, ALCo diversified into related product lines. In 1945, the company purchased the Beaumont Iron Works, based in Beaumont, Texas. This firm, which produced oil-field equipment, gave ALCo additional outlets for pipe, pressure vessels, and diesel engines. In the same year ALCo established the American Locomotive Export Company, a wholly owned subsidiary, to market its locomotives abroad, primarily in Latin America. As part of this diversification process, President Robert McColl appointed a new administrative committee to manage and coordinate the company's varied activities.[48]

Second, ALCo placed new emphasis on diesel locomotive development. The 1,500-hp RS-2, introduced in 1946, was the first successful road switcher to be offered by any builder. ALCo eventually sold more than three hundred RS-2 locomotives, along with more than twelve hundred of the similar, though slightly more powerful, RS-3. In 1946, ALCo introduced a 2,000-hp passenger locomotive (the PA) and a 1,500-hp freight locomotive (the FA). However, EMD's technologically superior GP-7, although not introduced until 1949, completely overwhelmed ALCo's pioneering RS-2. And, among other defects, PA locomotives often generated so much smoke (generally the result of an inadequate turbocharging system) that they violated some municipal antipollution ordinances.[49]

Furthermore, in the late 1940s ALCo delayed several years before making the extensive plant modifications that would be necessary for efficient diesel locomotive production. According to Board Chairman Fraser, "The company has been quick to reflect the demands of railroads for Diesel-electric locomotives . . . without disturbing our steam locomotive productive capacity. . . ."[50] Rather than abandoning its steam locomotive production facilities or converting them to other uses, ALCo mothballed them in anticipation of future orders, orders that never materialized.[51]

Between 1945 and 1947, ALCo invested approximately $20 million in facilities and equipment for diesel locomotive production. ALCo acquired the necessary funds by selling 46 percent of its holdings in the General Steel Castings Corporation, selling slightly less than half of its holdings in the Montreal Locomotive Works, assuming a $13.5 million bank loan, and issuing 400,000 shares of common stock.[52] In spite of management's claims to the contrary, this expenditure was not sufficient to establish "assembly-line production methods never before used in the locomotive-building field."[53] The physical arrangement of the Schenectady plant remained much as it had been during the halcyon days of steam locomotive production. As late as 1946, the plant was still cluttered with buildings (the boiler, tank, rod, jacket, and cylinder shops) that were best suited to steam locomotive production. The newly constructed diesel engine machine shop sat next to the flood-prone Mohawk River, on the opposite side of the 112-acre facility from the other diesel locomotive buildings. Materials, parts, and subassemblies certainly flowed less efficiently at Schenectady than at La Grange.[54]

ALCo's $20 million investment in new postwar production facilities, while impressive, was not nearly enough to keep pace with EMD. The tooling for EMD's diesel engine production alone represented a comparable investment. In addition, EMD had nearly twice that amount invested in its parts distribution network. La Grange itself was worth some $80 million, or four times the cost of ALCo's modernization program. Of course, ALCo could not hope to match the financial might of EMD's parent corporation; in 1960, ALCo's assets were 1 percent of GM's assets, its sales 0.7 percent of GM's sales, and its profits 0.005 percent of GM's profits. Much of EMD's expansion was financed through the division's own reinvested earnings, however, suggesting that GM's financial power did little to distort competitive patterns in the locomotive industry.[55]

ALCo's marketing and support services were likewise scantily developed. Not until 1948 did ALCo executives publicly indicate the importance of post-sales support services such as locomotive schools and parts distribution centers.[56] ALCo's locomotive school was never as extensive as EMD's and bore the additional burden of retraining thousands of ALCo's own employees, who had originally based their careers on steam locomotive technology.[57] As a result, ALCo's support services left much to be desired, causing one railroad executive to complain that ALCo's "service policies have been something terrible in the past. Our own experience in this respect has been most unfortunate...."[58]

ALCo's reliance on General Electric for critical locomotive components and for marketing expertise caused further difficulties. As with earlier units, ALCo locomotives used GE electrical equipment, and were thus marketed under an "ALCo-GE" nameplate. In order to boost the horsepower of some of its locomotives, ALCo used turbochargers that GE had originally designed for turbojet aircraft engines. As a result, the turbocharger operated at the unusually high speed of 10,300 revolutions per minute, and even the slightest manufacturing imperfection could cause a catastrophic turbocharger failure. To a certain extent, GE's products were thus too "high-tech" to mesh well with ALCo's more limited manufacturing capabilities.[59]

ALCo became more reliant on GE in 1950, when it transferred authority for diesel locomotive sales and service to that company—an arrangement that added more than three hundred GE sales outlets to ALCo's seven such facilities. ALCo had a national advertising campaign but allowed GE to prepare locomotive ads, since that company had better graphic facilities.[60] At the time, cooperation with GE benefited ALCo, since GE had greater strengths in advertising and marketing and since GE could operate far more sales and service outlets than ALCo could afford. The danger was that GE might decide to use its marketing capabilities as a springboard for entering the large diesel locomotive market on its own, without ALCo's participation. This possibility may well have occurred to ALCo executives, since GE was already heavily involved in the small diesel and export diesel markets. Given its limited technological and marketing expertise, ALCo had little choice in the matter. For ALCo, the collaborative benefits exceeded the risks, at least until 1960.[61]

Shrinking Demand

Although most American railroads did not complete their dieselization programs until the mid-1950s, ALCo's peak years of diesel locomotive production lasted only from 1946 to 1952. By 1953, ALCo had reduced its diesel locomotive output from four units per day to three. While 1951 locomotive sales were $131.2 million, those in 1954 were but $26.4 million.[62] In response, ALCo again altered its competitive strategy by diversifying into fields unrelated to railroading. ALCo's diversification program produced few financial benefits, largely because the company continued to manufacture customized products for the volatile capital goods industry.

The renaming of the American Locomotive Company as Alco Products, Incorporated (Alco) in 1955 symbolized the new diversification strategy.[63] The new company sold its holdings in the railroad-dependent General Steel Castings Company and purchased the Central Pipe Fabricating and Supply Company and the Carter Craft Company. Alco Products also merged with its wholly owned subsidiary, the Beaumont Iron Works, and established a new atomic energy products department.[64]

Alco offered a wide range of products in the railroad, atomic power, chemical, and petroleum industries.[65] In addition to diesel locomotives, these included diesel engines, diesel-powered drilling rigs, heat exchangers, pressure vessels, water heaters, steel pipe, cement kilns, evaporators, guided-missile components, steam generators, and some forty-four other products. Aside from its standardized, mass-produced diesel engines and locomotives, virtually all of these ancillary products, ranging from lock gates for the New York State Barge Canal to shields for the Lincoln Tunnel, were custom-engineered and custom-built. Because of intense competition, profit margins on these products remained low, and the cyclical nature of their target markets (oil, defense, public works) actually exacerbated the "feast or famine" tendencies associated with the locomotive industry. Alco neither reduced its dependency on locomotive orders nor carved out niche markets that were isolated from competition by larger firms, primarily because Alco chose the wrong products in the wrong industries.[66]

In 1956, Alco introduced a new line of locomotives, including the 900-hp S-6 switcher and the 1,800-hp RS-11 road switcher. Alco's Model 251 diesel engine constituted the primary advantage of these models. Although the 251 offered improved performance and reliability over the older Model 244, Alco timed its introduction poorly. Railroad executives who had just purchased older Alco locomotives during the initial dieselization boom were disgruntled that their nearly new locomotives embodied outdated technology. The Model 244 and the Model 251 had few parts in common, obligating railroads to handle many additional replacement parts, and forcing Alco to make extensive modifications to its manufacturing equipment and techniques.[67]

To produce its new line of diesels more efficiently, Alco initiated a major plant modification program in 1957. The company reduced the size of its Schenectady plant by a third, mostly by selling or demolishing vacant buildings formerly used for combat tank and steam locomotive production. Alco's latest physical plant modernization program, enacted some twenty-two years after comparable developments at EMD, created a "progressive station" assembly line in which production bays on either side of the main erecting shop funneled components onto one of three assembly tracks. This system speeded production and reduced the costly transfer of locomotive components between various parts of the plant.[68] These enhancements came too late to salvage Alco's position in the domestic diesel locomotive market, since, by 1957, the dieselization boom had ended.

Unfortunately for Alco, its diesels did not live up to expectations. One railroad, the Northern Pacific, believed that Alco locomotives performed as well as EMD units, but with much higher maintenance costs. Maintenance difficulties arose, in part, from the need to remove the locomotive engine's cylinder heads and pistons in order to perform even routine inspections—something not necessary on EMD locomotives, which had been designed to allow visual inspection of cylinder liners, pistons, fuel injectors, piston rings, and other critical parts.[69] Overall, Northern Pacific officials felt that the basic problem arose "from trying to squeeze too many horsepower in a four-cycle engine into too small a space. . . ."[70]

Despite the limitations inherent in Alco locomotives, the Northern Pacific executives, like those at many other railroads, continued to purchase that company's products because they feared that an EMD monopoly would stifle technological progress and artificially increase locomotive prices. Railroad correspondence contained references such as "The principal reason advanced for purchasing Alco locomotives is to assure more than one manufacturer remaining in business so as to avoid dangers inherent in a monopoly,"[71] and "I think it is desirable to keep at least one good competitor in the field with EMD."[72] One of its customers actually believed that "the sales policy of the American Locomotive Company has been that the railroad industry should take an inferior product just to keep a second manufacturer alive. . . ."[73] Alco's willingness to lean on the antimonopoly crutch did little to encourage customer loyalty and would leave the company vulnerable if a more efficient producer were to enter the diesel locomotive industry.

EMD and ALCo in the Canadian Market

Canada offered a lucrative foreign market for EMD and the other U.S. locomotive producers. Although Canadian railroads embraced dieselization at a later date than their American counterparts, diesels sold in Canada were virtually identical to those produced for sale in the United States. This was not simply an export market, however, since Canadian domestic-content laws mandated that a large portion of the locomotive be manufactured in that country. While Canadian railroads purchased experimental diesel locomotives as early as 1922, the dieselization of Canadian railways did not become profitable until railroad traffic increased substantially in the late 1940s and early 1950s. Reversing the typical U.S. dieselization pattern, Canadian railroads assigned diesels to freight service before employing them on passenger runs, partly out of fears that diesels lacked sufficient steam-heating capacity for wintertime passenger trains. In 1947, oil fields in northern Alberta began production, thus reducing the cost differential between diesel fuel and steam locomotive coal. By 1952, more than 610 diesels were in use on Canadian railroads, approximately 10 percent of the locomotives then in service in Canada. Significantly, these were purchased to accommodate increased traffic, not to replace existing steam locomotives.[74]

The organizational strengths that gave EMD control over America's locomotive industry enabled the division to capture a substantial share of the Canadian market as well. EMD's dominance was not complete, however, and ALCo ultimately enjoyed more success in Canada than in the United States.[75] To comply with Canadian domestic content laws, GM established a subsidiary, GM Diesel, Limited, to assemble and market diesel locomotives. This was part of a larger GM expansion program in Canada that included an increase in Canadian automobile production and the construction of a 500,000-square-foot plant for Frigidaire Products of Canada. In June 1950, GM Diesel completed a 226,000-square-foot plant in London, Ontario, that would employ one thousand people and have a capacity of one locomotive per day. EMD manufactured diesel engines, generators, and traction motors at La Grange, while GM Diesel fabricated locomotive bodies, underframes, trucks, and ancillary electrical equipment and then assembled the final product. By January 1952, GM Diesel had built more than two hundred locomotives in Canada.[76]

Because of its investments in manufacturing in Canada, EMD and GM Diesel captured much of the $500 million steam locomotive replacement market in Canada. However, GM Diesel's success in Canada never approached the level achieved by EMD in the United States and ALCo-designed locomotives continued to offer stiff competition to EMD designs long after ALCo itself had ceased diesel locomotive production. The two major Canadian railroads, the Canadian National and the Canadian Pacific, often preferred ALCo designs, which were better-suited to the nation's severe climactic variations. In addition, ALCo's subsidiary in Canada, the Montreal Locomotive Works, had a much longer tradition of Canadian operations (its predecessor, the Locomotive and Machine Company of Montreal, Ltd., opened in 1902), and seemed to many Canadian customers to be a more "Canadian" company than GM Diesel and less obviously a subsidiary of a U.S. corporate colossus. Finally, EMD had such a massive backlog of orders, especially during the 1950s, that it devoted relatively little effort to foreign markets, including those in Canada, until the 1960s, by which time ALCo/MLW had captured a substantial lead in the Canadian market.[77]

ALCo was the first U.S. company to build diesel locomotives in Canada. In 1948, MLW completed the first production-model diesel locomotive built in Canada, although the company continued to manufacture steam locomotives until 1950. At first, MLW merely assembled diesel locomotives, using ALCo engines imported from the United States and electrical equipment supplied by the Canadian General Electric Company's plant in Peterborough, Ontario. The following year ALCo agreed to license the Dominion Engineering Works to produce ALCo engines in Canada for use in MLW locomotives.[78]

ALCo received little benefit from MLW's success, since it had sold three-fourths of its stock in MLW in April 1946, in order to pay for its own modernization program at Schenectady. MLW prospered during the 1950s and 1960s, even as ALCo's fortunes declined. The Canadian company continued to manufacture diesel locomotives and related railroad equipment for the Canadian and non-U.S. export markets. When ALCo ended U.S. locomotive production in 1969, the company transferred its diesel engine and locomotive designs to MLW. Canadian-owned Bombardier, Ltd., a struggling producer of snowmobiles, acquired MLW in 1975; and, although diesel locomotive production has ceased, the company still sells mass-transit equipment and other railroad related products.[79]

Industry Maturation and Shakeout: The Demise of Baldwin, Lima, and Fairbanks-Morse

As ALCo and EMD vied for market dominance in the decade after World War II, Baldwin, Lima, and Fairbanks-Morse battled for survival. During the postwar rush to dieselize, they eked out an existence by relying on traditionally loyal customers from their days as steam locomotive producers (in the case of Baldwin and Lima), by taking advantage of railroad policy decisions to purchase some locomotives from each builder before concentrating on a narrower range of models, and by taking orders that could not be accommodated at La Grange or Schenectady. By 1955, however, as railroads replaced the last of their steam locomotives, these marginal producers had either left the industry or were about to do so. While their problems varied, in general, all these companies seriously underestimated the technological, managerial, and financial resources necessary to compete successfully against ALCo and EMD. Executives at both Baldwin and Lima realized just how difficult it was to transform their companies from small-batch producers of customized steam locomotives to large-batch manufacturers of standardized diesel locomotives. Fairbanks-Morse realized that competency in both the railroad equipment and diesel engine fields did not guarantee success in the diesel locomotive industry.[80]

Baldwin's Collapse

Baldwin's woes intensified in the postwar period. As a result of reductions in steam locomotive and ordnance production, Baldwin's sales declined from a high of $221.5 million in 1943 to $85.3 million in 1946. Partly because 1947 steam locomotive production at Baldwin still exceeded that of diesels, 227 to 163, the company was not completely convinced that steam locomotives were a lost cause.[81] In 1946, Charles Kerr, consulting transportation engineer at Baldwin's production partner, Westinghouse, predicted that "Better steam locomotives than we have ever known before are on their way," and concluded that "the old 'iron horse' is far from dead."[82] A year later, a Baldwin spokesman still believed that "the Diesel locomotive is in its infancy...."[83] Not until 1948 did Baldwin determine that "it seems evident that future railroad motive power purchases for use in this country will be chiefly of the Diesel type and that further large [steam] export orders must be considered as highly uncertain."[84]

Some of these public statements of support for steam locomotive technology may have represented efforts by Baldwin executives to exonerate themselves in the eyes of their presumably displeased shareholders. However, Baldwin executives expressed similar opinions before diesels constituted viable technology, before EMC had attained market dominance, while EMD was losing market share during World War II, and as steam locomotive orders evaporated during the early postwar period. As such, Baldwin executives appear to have been unusually predisposed to deceive their stockholders, their customers, and the general public; or, more likely, a significant proportion of Baldwin executives really did believe that diesels would not be able to drive steam locomotives from American railroads. Even as late as 1955, a Baldwin vice president testified before a Senate subcommittee that "I wouldn't say we have abandoned the [steam locomotive] field, because that is the last of our intentions."[85]

The informal, customer-driven production and marketing requirements of the steam locomotive industry had worked well for many decades and caused Baldwin to develop a strong—and ultimately counterproductive—loyalty to its customers. Since some railroad executives were themselves loyal to steam locomotive technology, Baldwin managers felt that they had little choice but to manufacture what their customers demanded. As a Baldwin executive admitted in 1955, "We could not drop the steam locomotives and concentrate all our efforts on the diesels because there were several railroads . . . that still stuck with the steam locomotive, they were our customers, and we had an obligation to continue production for their use."[86] The customer-driven character of the steam locomotive industry meant that

Baldwin could not abandon steam locomotive technology, even when it was in the firm's best interest to do so.

Unlike ALCo, Baldwin had not placed much emphasis on diesel locomotive research and development during the war years, even though the WPB had placed no restrictions on these activities. As such, Baldwin entered the postwar era well behind ALCo, and even further behind EMD. In 1946, Baldwin introduced a 1,500-hp road switcher, the DRS-6-4-15, contemporaneous with the introduction of ALCo's new postwar road switcher and three years before the debut of EMD's GP-7. In spite of this early lead, Baldwin sold only a few dozen of this model, probably because of the company's reputation for poor quality.

While Electro-Motive had first penetrated the road freight locomotive market (with the FT, in 1941), and had only later begun road switcher production, Baldwin followed the reverse course. According to Baldwin, "The year 1947 was one of transition in the locomotive department during which [the] Company undertook its first substantial program in the construction of large diesel road locomotives and, naturally, experienced many of the difficulties which occur in the establishment of a new project."[87]

Baldwin introduced new diesel locomotive designs three years later, claiming that these locomotives featured increased horsepower, easier maintenance, and a high degree of interchangeability between parts and subassemblies. Interchangeability had its limits, however, since Baldwin still struggled to standardize production methods.[88] When one Baldwin executive conducted an "investigation [that] brought out the fact that a great number of the [Model] 600 engine parts are common for the following types: Locomotive, Stationary, and Marine, as well as the new [Model] 700 design," a company engineer penciled in "NO—only a few minor parts from 600 series engines are used on 700."[89] Apparently, Baldwin exhibited a certain degree of miscommunication among its executives. The company's procedure for coordinating various design elements and then translating these paper designs into three-dimensional products "proved very unsuccessful due to the separation of detail draftsmen from the designers and lay-out men." A Baldwin analysis of the situation accordingly recommended that *"the design group should therefore have the detailers work in the same room and under the immediate supervision of the designers and lay-out men."*[90] Weekly meetings of the newly established Diesel Engine Cost Reduction and Production Committee during 1948 and 1949 did little to alleviate these problems.[91]

As had been the case during World War II, Baldwin engineers and executives frequently paid scant attention to market conditions as they designed these locomotives. They pondered whether we [should] design the best product possible and, after determining the cost thereof, establish a sales price for the product which can be sold on its merits regardless of price, or should we analyze our markets to determine the price at which the product can be sold successfully and then engineer the product to suit the price? . . . it would be very helpful if the Engineering Department could get some idea of the restrictions that are placed on us with respect to this engine by virtue of price limitations as dictated by market conditions.[92]

In the heyday of the steam locomotive industry, railroads and builders determined price, based on a unique set of performance criteria, as much through negotiation as through competitive bidding. Market research accordingly had limited value, and as a diesel locomotive producer, Baldwin struggled to master this new skill.

Baldwin's post-sales support services lagged far behind those offered by its two larger competitors. Baldwin did not establish a nationwide parts distribution network and frequently relied on Westinghouse to provide this service. In addition, Baldwin's locomotive school was of dubious quality. Although Westinghouse offered classes at both East Pittsburgh and Eddystone, beginning in 1942, Baldwin's own locomotive school did not open until 1948. Baldwin's twenty-hour course used "life-size operating panel boards" to simulate actual locomotives at a time when EMD was sending real locomotives on training duties around the United States. By the late 1940s, furthermore, railroads themselves were beginning to establish their own educational programs, thus rendering the training services provided by builders in the 1930s and early 1940s less valuable.[93] Baldwin's late arrival in the locomotive market also reduced the effectiveness of the company's spare parts repair and replacement services. As one railroad executive noted, these programs were of greatest value to railroads just beginning to replace their steam locomotives with diesels. They had correspondingly less value by the 1950s when carriers had developed their own maintenance capabilities.[94] Baldwin's difficulties in establishing warranty terms with an outside supplier of vitally important turbochargers added to these problems.[95]

Westinghouse Increases Its Control over Baldwin

In an attempt to protect its investment in Baldwin and maintain its market for locomotive electrical equipment, Westinghouse increased substantially its control over Baldwin in 1948. In July of that year, Westinghouse purchased 500,000 shares of authorized but unissued Baldwin common stock for $15.11 per share. This purchase, combined with earlier Westinghouse holdings, gave that company ownership of 22 percent of the 2,375,553 shares of outstanding Baldwin stock. In addition to providing Baldwin with $7.5 million in desperately needed working capital, the purchase also gave Westinghouse a continued incentive to supply electrical equipment.[96]

Westinghouse then sent several of its own managers to Baldwin in an attempt to halt that company's decline. Marvin W. Smith, a former vice president in charge of engineering at Westinghouse, became Baldwin's new executive vice president and chief executive officer. James R. Weaver left his position as the head of manufacturing at the Westinghouse plant in Springfield, Massachusetts, to become vice president of manufacturing at Eddystone. John S. Newton, an expert in steam and gas turbines at Westinghouse, became vice president of engineering at Eddystone. Ralph Kelly, who at the time of the stock sale had been both president of Baldwin and executive vice president at Westinghouse, resigned in order "to be afforded greater freedom for personal and civic activities," although he continued as chairman of the board at Baldwin.[97]

Befitting their increased control, Westinghouse made its opinions known to Baldwin managers. When Baldwin suggested that it send terminal boards to East Pittsburgh for pre-wiring, Westinghouse issued an immediate rejection, stressing that this procedure would not mesh well with its own operational routines. Westinghouse also made unilateral design changes in its generators and other electrical equipment, assuring Baldwin that these alterations would benefit Baldwin's customers. Eventually, by January 1, 1953, Westinghouse assumed total control over the repair and renewal of all electrical equipment used on Baldwin locomotives.[98]

The increased involvement of Westinghouse did little to benefit Baldwin. The company experienced a substantial decline in business during 1949, with the result that it reduced its workforce to 9,587 by the end of the year—less than half the wartime high of 20,095. A 1951 article in *The Magazine of Wall Street* criticized Baldwin's poor management and its inability to adopt modern manufacturing methods, concluding that "no matter how favorable business conditions may be, there always seem to be some companies which fail to gain their rightful share of prosperity."[99] In the years that followed, this "rightful share of prosperity" became even more elusive. Sales of locomotives and parts, which had been 45 percent of all business in 1951, declined to 30 percent in the following year.[100]

This rapid decline in diesel locomotive sales prompted Westinghouse to sever its financial ties with Baldwin. In early 1954, Westinghouse announced that it was withdrawing from the locomotive electrical equipment market, leaving GE as the only independent supplier of these products. Westinghouse continued to manufacture equipment for urban mass transit systems and similar applications. In May 1954, Baldwin repurchased 515,000 shares of its own stock from Westinghouse, at a loss to the latter company of $6.11 per share.[101]

Diversification and Departure

Baldwin attempted to rectify its continued difficulties in the capital equipment sector by continuing its established policy of diversification and by starting a new policy of merger and decentralization. The highlight of this program was a November 1950 merger with the Lima-Hamilton Corporation, an event described in greater detail later in this chapter. Briefly, Baldwin exchanged 1.9 million shares of its stock for the stock of its smaller competitor on a share-for-share basis. In order to make the arrangement as lucrative for Baldwin stockholders as possible, the company transferred the asset-rich Midvale Company and General Steel Castings Corporation, along with the proceeds from the earlier sale of the Flannery Bolt Company, to the newly created Baldwin Securities Corporation. Baldwin shareholders then received the stock of the Baldwin Securities Corporation as a dividend. While this payment provided a substantial bonus for shareholders, the removal of these assets reduced the ability of the new Baldwin-Lima-Hamilton Corporation to consolidate its production and distribution facilities.[102]

Baldwin-Lima-Hamilton also created a multidivisional structure, with a "centralized executive policy and decentralized administration"—much like GM's successful organizational form, albeit on a much smaller scale.[103] To a certain extent, this new strategy worked. In 1957, Baldwin-Lima-Hamilton was featured in "The Security I Like Best" column of *The Commercial and Financial Chronicle*, which said that the company had "definitely turned the corner. . . ."[104]

However, this corporate rebirth involved a final realization that Baldwin could not continue to participate in the diesel locomotive industry. Baldwin built its first diesel locomotive in 1925, and its last in 1956. The company rarely, if ever, earned a profit on its diesel locomotives and between 1946 and 1952 lost money every year on combined (steam and diesel) locomotive sales. By 1954, Baldwin-Lima-Hamilton executives had finally reached the inescapable conclusion that they should leave the industry. By then, railroads had standardized on EMD or ALCo locomotives and Baldwin would have found few buyers, even with adequate locomotive designs.

Baldwin-Lima-Hamilton soon began to dispose of the vast buildings originally designed for steam locomotive production but poorly adapted to the manufacture of diesels. By 1956, 75 percent of Eddystone—300 of the original 381 acres, including eight buildings—had been sold. Two years later, virtually all of the former locomotive facilities at Eddystone had been demolished.[105]

Baldwin frittered away chances for success in the diesel locomotive market. In the 1920s and 1930s, Baldwin entered and remained in that market to provide diesels as a favor for longtime Baldwin steam locomotive customers and to explore the possibilities of what Baldwin executives believed was a niche market. By artificially reducing the output of EMD, the War Production Board gave Baldwin a small captive market during World War II. After the war, the insatiable demand for diesels kept this inefficient, marginal producer in business. Once the postwar dieselization rush had subsided, railroads had little reason to purchase Baldwin's inadequate products.

One individual in particular exacerbated Baldwin's difficulties. Samuel Vauclain, who was as closely associated with Baldwin as Henry Ford was with the Ford Motor Company, reached the pinnacle of his career at the same time that Baldwin came to dominate the steam locomotive industry. Vauclain could not conceive of a world without Baldwin steam locomotives, any more than Henry Ford could imagine a world without the Model T. As such, Vauclain, along with other like-minded Baldwin executives, had little desire or incentive to plan for the ultimate replacement of steam locomotives by diesels.

To some extent chance—or at least bad timing—played a role in Baldwin's demise. Baldwin began construction of its massive new Eddystone facility during the first decade of the twentieth century just as, unbeknownst to its executives, the company reached the pinnacle of its success.[106] Baldwin completed its move to Eddystone in 1929, just prior to the unanticipated onset of the Great Depression. The Eddystone albatross subsequently crippled Baldwin financially. In the end, the company that had been one of the world's most respected producers of steam locomotives for more than a century failed in the diesel locomotive industry.

Lima: A Truly Marginal Producer

During the decade following World War II the Lima Locomotive Works, the least successful of the six producers that are the subject of this study, entered and then rapidly left the diesel locomotive industry. From the 1930s until its merger with Baldwin in 1950, Lima struggled to cope with the decline of its established heritage of custom-craft production and with an evolving industry that was becoming increasingly dominated by larger firms. More than any of its competitors, Lima lacked the financial resources necessary to create effective organizational capabilities in the diesel locomotive industry.

The Lima Locomotive Works enjoyed considerable success in the steam locomotive industry, thanks largely to the company's control of specialized technology, in the form of the patented Shay locomotive. In 1869, Carnes, Harper, and Company began building threshing machines, portable sawmills, and stationary steam engines. Its owners reincorporated the company as the Lima Machine Works in 1877, just prior to a decrease in thresher and sawmill demand in the late 1870s. The company accordingly approved, in 1880, a proposal by Ephriam Shay, a Michigan lumberman, to build an experimental locomotive for use in the logging industry. This type of steam locomotive, which Shay patented in 1881, used a geared drive system to negotiate steeper grades and sharper curves than conventional rod locomotives could accommodate. Lima acquired all patent rights to the Shay and produced 2,761 of these engines between 1880 and 1945. Lima sold most of its Shay output to shortline logging and quarrying operations, and as a result, the company maintained fewer contacts with mainline railroads than did ALCo or Baldwin.[107]

This situation changed in 1922, when Lima began the development of radically improved mainline steam locomotive designs. Under the direction of W. E. Woodward, vice president in charge of engineering, Lima's engineers made dozens of improvements to conventional steam locomotive designs, greatly increasing efficiency and lowering fuel consumption. The Lima A-1 "Superpower" locomotive made its debut in 1925 and was well received by many railroads. ALCo and Baldwin quickly copied many elements of "Superpower" technology, which remained state of the art until the end of steam locomotive production in the United States.[108] During the 1920s and 1930s, Lima made numerous other smaller contributions to steam locomotive technology.[109]

In spite of Lima's commitment to improving steam locomotive technology, the company did experiment with diesel locomotives. Lima built two small gasoline-mechanical locomotives in 1909 or 1910. These were suitable only for light industrial switching and were not produced in quantity. Lima manufactured similar experimental models in 1929 and for a few years thereafter, although the Great Depression ended any plans the company may have had to continue production. The company also showed interest in diesel locomotive developments in England during the early 1930s. Lima, still actively involved in improving steam locomotives and feeling the effects of the depression, chose not to enter the diesel locomotive market—a wise decision, since the small company lacked the capital necessary to develop or purchase adequate diesel locomotive technology.[110]

Lima developed little knowledge of diesel locomotive technology during either the depression or World War II.[111] The enormous increase in steam locomotive orders during the war (Lima built more than a thousand of these) represented the last gasp of a once mighty industry. However, Lima officials may have misinterpreted this temporary increase as an indication of a large postwar market for steam locomotives. They continued to improve steam locomotive technology for several years after the war, and President John Dixon initially maintained that "the modern steam locomotive [was] far from dead or defeated."[112]

By early 1947, however, Lima officials realized that domestic demand for steam locomotives had ceased. In its 1949 annual report, Lima acknowledged that "there is almost no demand currently for steam locomotives at home."[113] Lima executives accordingly decided to follow the earlier lead of ALCo and Baldwin and enter the diesel locomotive market.

Since Lima possessed little competence in diesel engine technology, the company merged with the General Machinery Corporation of Hamilton, Ohio, in October 1947. General Machinery had produced its first diesel engine (for maritime use) in 1924, and several of its engines had been installed in self-propelled railroad cars. General Machinery produced a single experimental diesel-hydraulic locomotive in 1939. By 1947, the General Machinery Corporation produced steam, diesel, and gasoline engines, machine tools, machinery for can-making and sugar-cane milling, welding equipment and supplies, and various types of hydraulic presses.[114] Following the merger, George A. Rentschler, the former president of the General Machinery Corporation, became chairman of the executive committee at the new Lima-Hamilton Corporation.[115] Rentschler, appalled by the inefficiency, mismanagement, and waste at the Lima facility, soon undertook a number of cost-saving measures, ranging from banishing photographs from the annual reports to firing hundreds of Lima employees.[116]

In spite of Rentschler's economy measures, Lima-Hamilton did not integrate its production facilities. The company continued to build diesel engines in Hamilton and ship them north by rail to Lima, where workers installed them in the locomotive shell. In addition, the company purchased generators and traction motors from Westinghouse.

Lima-Hamilton executives decided to concentrate initially on the production of 1,000-hp to 1,500-hp switch engines, thus entering an already oversaturated market. ALCo, Baldwin, Fairbanks-Morse, and EMD all produced similar products. Electro-Motive, in particular, had spent more than a decade refining its switcher line. Furthermore, by the late 1940s, most railroads had already replaced their highly inefficient steam switchers with diesels and were instead beginning to purchase more powerful and versatile road locomotives for long-distance freight and passenger service.[117]

Lima-Hamilton tested its first diesel engines in July 1948, and completed its first production diesel locomotive ten months later. Still, the company remained temporarily committed to steam locomotive production. During 1948, the company produced and delivered thirty-six steam locomotives and had thirty-one more under construction at the end of the year. These were delivered in 1949 and had the distinction of being the last steam locomotives produced in the United States for domestic use.[118]

By 1950, Lima-Hamilton's position in the increasingly competitive diesel locomotive market was tenuous at best. Fairbanks-Morse announced a new line of standardized locomotives that embodied numerous technical improvements. Baldwin unveiled its own improved and standardized line during the second half of 1950. Baldwin offered 800-hp, 1,200-hp, 1,600-hp, and 2,400-hp units that were virtually identical to Lima-Hamilton models, but with improved performance. And, in October 1949, EMD introduced its extremely popular GP-7 road switcher.[119]

Accordingly, in August 1950, the boards of directors of Lima-Hamilton and Baldwin announced plans for a merger of their two companies, claiming that "in the Diesel-electric field there should be substantial advantages in integrating the activities of the two organizations." On October 25, 1950, stockholders approved the creation of the new Baldwin-Lima-Hamilton Corporation. The new corporation issued an additional two million shares of common stock, most of which was exchanged for Lima-Hamilton stock on a one-for-one basis. Marvin W. Smith, the former president of Baldwin, became the president of Baldwin-Lima-Hamilton, and George Rentschler was elected chairman of the board. His nephew, Walter Rentschler, became vice president in charge of the Lima-Hamilton division.[120]

After absorbing its smaller rival, Baldwin terminated locomotive production at Lima. In August 1951, Lima-Hamilton completed the last of 7,823 locomotives at its Lima facility, only 174 of which were diesels.[121] Since Lima-Hamilton locomotives were virtually identical to their Baldwin counterparts, the manufacturing rationalizations that followed the merger included the discontinuance of the original Lima-Hamilton locomotive line. Westinghouse purchased the plans and designs for these Lima-Hamilton locomotives but chose not to produce them.[122]

A number of factors contributed to Lima's inability to transfer its skills as an innovative and successful producer of steam locomotives to the production of diesel locomotives. Lima entered the diesel locomotive market too late to be an effective competitor. The company lacked the technical knowledge and financial resources necessary to develop its own diesel engines and electrical equipment. Lima built only the frame and superstructure of the locomotive, portions which were the least technically sophisticated. This limited familiarity with diesel engine technology forced Lima into an unfavorable merger with a company (General Machinery Corporation) that produced a diesel engine unsuited to railroad applications. When Lima began diesel locomotive production, it chose to manufacture 800-hp to 2,400-hp switchers. This virtually guaranteed failure, since that market niche was already filled. If Lima had instead produced small industrial units, or custom-built locomotives for export, it might have been more successful. Lima's slow pace of production (it built as many diesels in two years as EMD could produce in six weeks) prevented the company from amortizing its research and development expenditures.[123] Finally, Lima-Hamilton did not offer the after-sale support services—parts depots, repair and rebuilding facilities, or diesel locomotive training schools—that were vital to successful diesel operation, customer satisfaction, and repeat orders.

Ultimately, however, Lima's difficulties stemmed from the company's small size and limited capitalization. Financial limitations prevented the company from acquiring adequate technological capabilities or from rationalizing and modernizing its production facilities. Since Lima's total capitalization in 1945 was less than half of Electro-Motive's diesel engine R and D expenditures during the decade of the 1930s, it is not difficult to understand why Lima and Lima-Hamilton experienced difficulties. Given its financial limitations, Lima-Hamilton did well to produce as many diesel locomotives as it did.

Fairbanks-Morse in the Postwar Locomotive Market

By 1940, Fairbanks-Morse executives reasoned that they could combine their long-established contacts in the railway equipment supply industry with their impressive knowledge of diesel engine development and production. They overestimated the value of their organizational capabilities, however, and considerably underestimated the difficulty of translating skills in diesel engine production into success in the diesel locomotive market. Company executives evinced little familiarity with their company's investments in the locomotive industry.[124] Growing divisions within the ranks of Fairbanks-Morse's family management compounded these problems and ultimately destroyed the company.

Diesel engine production, primarily for submarine applications, soared during the war, thanks in part to a massive government-financed expansion of Fairbanks-Morse's Beloit manufacturing facility. As company executives had feared, Fairbanks-Morse experienced a serious decline in revenue following the end of World War II. Net sales fell from a 1943 high of $183.8 million to $56.6 million in 1946. Profits increased during that time, however, from $2.8 million to $3.1 million.[125]

Based on the unprecedented postwar demand for diesel locomotives, Fairbanks-Morse planned to expand its wartime locomotive line by offering road freight and road passenger locomotives, in addition to yard switchers, first offered in 1944. In November 1946, Fairbanks-Morse announced its decision to expand its locomotive production facilities at Beloit for this purpose. Fairbanks-Morse also planned to spend $5.2 million to acquire and modify the government-owned buildings at Beloit, in order to manufacture a full range of diesel locomotives, ranging from 1,000 hp to 8,000 hp.[126]

Because Fairbanks-Morse wanted to introduce its new diesel line as soon as possible, it chose to subcontract the assembly of its diesel locomotives until its new facilities at Beloit were completed, and this policy led to one of the most unusual competitive patterns in American industry. Fairbanks-Morse awarded the assembly contract to GE, a company that simultaneously built electrical equipment for one of Fairbanks-Morse's competitors, ALCo. Fairbanks-Morse originally designed its regular production locomotives to accommodate electrical equipment manufactured by GE's competitor, Westinghouse. Westinghouse in turn supplied electrical equipment to Baldwin, another competitor of Fairbanks-Morse. GE built 111 locomotives in this manner before the expanded Beloit plant was ready in 1948, yet the haphazard nature of this production alliance undoubtedly hurt the efforts of Fairbanks-Morse to compete against EMD, which had integrated its manufacturing capabilities a decade earlier.[127]

Fairbanks-Morse lacked both suitable production facilities and successful locomotive component designs. The company had not designed its Beloit facilities specifically for diesel locomotive production, and the plant was never really adequate for that purpose. In addition, its opposed-piston engine design, although well suited for submarines, performed poorly in locomotive applications. Dirt and vibrations, an inescapable part of railroad freight and passenger service, proved more deleterious than the rigors of the sea. Excessive consumption of lubricating oil increased operating costs. Fairbanks-Morse locomotives spent an inordinate amount of time in the shops, and repair personnel found it difficult to obtain access to key parts and subassemblies. Cylinder liner replacement caused especially serious difficulties. Fairbanks-Morse received "several reports of engines stopping on the road and the crew being unable to start them again."[128] These problems eroded confidence in Fairbanks-Morse locomotives.

Fairbanks-Morse did not establish a successful marketing structure, and this omission reinforced the belief of many railroad executives that Fairbanks-Morse produced unreliable locomotives. Company sales efforts were particularly amateurish. Executives encouraged their sales forces to think of themselves as "lumberjacks" who could garner so many "board feet of lumber," as pilots in an "airplane race," as cowboys attempting to "laso the most cattle," or, during World War II, as submarine captains out to get the most "enemy ship sinkings." If this were not enough to motivate salesmen, Fairbanks-Morse also sponsored the annual "Big Turkey Contest"—appropriately named, given the company's performance in the locomotive industry.[129] Fairbanks-Morse established regional sales and service offices, but these were neither widespread nor well supplied. Likewise, railroads rarely patronized Fairbanks-Morse's diesel locomotive training programs.[130]

The cost of these marketing weaknesses is illustrated by Fairbanks-Morse's relationship with one of its best customers, the Southern Pacific. In 1953 and 1954, the Southern Pacific purchased sixteen Fairbanks-Morse Model H24–66 "Trainmaster" locomotives, which it assigned to service on lines in New Mexico and West Texas. The locomotives performed poorly there and experienced chronic maintenance difficulties. This situation persisted until the railroad transferred most Fairbanks-Morse locomotives to low-status commuter service in the San Francisco area. Local Southern Pacific shop forces gradually developed an understanding of the locomotives' idiosyncrasies, and reliability and performance improved dramatically. Had Fairbanks-Morse aggressively implemented proactive post-sales training programs for railroad maintenance employees, it might have forestalled many of the problems that were commonly, if inaccurately, blamed on poor locomotive design.[131]

Family Problems

Managerial difficulties further reduced Fairbanks-Morse's effectiveness in the diesel locomotive industry. Far more than any of its competitors, family management imposed a burden on Fairbanks-Morse. Other companies did exhibit familial ties—Charles Kettering placed his son, Eugene, in charge of diesel engine development efforts at Winton, and Samuel Vauclain's son served as a vice president of Baldwin, until advancing age and ill health forced him to resign (his father continued working). Nowhere else, however, did nepotism reach the heights that it did at Fairbanks-Morse. Charles Hosmer Morse joined the company in 1850. His son, Charles H. Morse Jr., served as president from 1914 until 1927, when he became chairman of the board. In 1927 his brother, Col. Robert H. Morse, was elected vice chairman of the board. Born in 1878, Robert Morse began as an apprentice at Beloit in 1895 and, not surprisingly, was soon promoted to salesman, department manager, sales manager, and, finally, president of the manufacturing division. He became a Fairbanks-Morse vice president in 1918. By 1945, he was president and general manager and a director (at the same time). Charles H. Morse Jr. still occupied a seat on the board of directors at this time, as did his nephew Charles H. Morse III (son of Col. Robert Morse). C. H. Morse III was also president of Inland Utilities, a Fairbanks-Morse subsidiary, and later became vice president in charge of manufacturing. At the same time, Robert H. Morse Jr., son of Charles H. Morse Jr., was vice president, general sales manager, and a member of the board of directors. A fifth Morse, John, was coming up through the ranks as the assistant manager of Fairbanks-Morse's San Francisco Branch when his career was cut short by a fatal auto accident.[132]

As late as 1957, the Morses owned more than a third of Fairbanks-Morse's stock, and these managers were more than willing to pay substantial dividends to the principal stockholders (themselves). The company made little investment in research and development or in plant and equipment. Instead, according to one financial analyst, "the Morse family reputedly ran the company like a private preserve."[133] This type of family mismanagement, perhaps more typical of British industry, put Fairbanks-Morse at a serious competitive disadvantage with companies, such as GM, that had been able to separate ownership from management.[134]

Between 1954 and 1958, a takeover bid by corporate raider Leopold Silberstein's Penn-Texas Corporation bitterly divided the Morse family.[135] The Penn-Texas Corporation took advantage of this situation by purchasing 77 percent of Fairbanks-Morse stock and, in November 1958, merged with the latter firm to create the Fairbanks-Whitney Corporation. This company was soon in serious financial trouble. Executive salaries remained excessive, production was inefficient, and the sales force failed to secure important contracts. During the first quarter of 1958, net earnings declined 97 percent from the same period a year earlier. Between 1955 and 1963, stock prices decreased from almost $23 a share to $4 a share. In 1962, the company lost $4.8 million. In May 1962, George A. Strichman became president of Fairbanks-Whitney, declaring unhelpfully, "This is the worst company I've ever seen."[136] In 1963, Fairbanks-Whitney became the Fairbanks-Morse subsidiary of Colt Industries.[137]

By 1957, locomotive sales at Fairbanks-Morse had declined to virtually nil. Regular locomotive production ended two years later, although Fairbanks-Whitney built a few units for railroads in Mexico until 1963. Thereafter, the company continued to manufacture opposed-piston engines for marine and stationary use. In all, Fairbanks-Morse built or supervised the construction of 1,256 locomotives, 1,100 of which were built during the period of peak production between 1944 and 1954.[138] On average, locomotive sales accounted for only about 15–20 percent of total company revenues.[139] The company's market share was never large—4 percent in 1949 and 6 percent in 1950, for example. Fairbanks-Morse made a profit on locomotive sales only in 1948, when it earned 4 percent. Two years later, it incurred a 14 percent loss on sales. During the 1950s, the rate of return on Fairbanks-Morse's investment in the locomotive industry was negative 14 percent.[140]

Like Lima, Fairbanks-Morse never really had a chance in the diesel locomotive industry. The company derived little benefit from its contacts in the railway equipment supply industry and found that its opposed-piston engines, although sophisticated and well-built, were ill-suited to locomotive applications. Fairbanks-Morse entered the market too late to compete against established firms like EMD. The company lacked the financial resources to develop an adequate marketing program, particularly in the realm of post-sale support services. The family management at Fairbanks-Morse compounded these difficulties.[141]

Common Difficulties

In retrospect, Baldwin, Lima, and Fairbanks-Morse had few opportunities for long-term success in the diesel locomotive industry. While certain factors were beyond executive control (i.e., poor timing at Baldwin), managers were at least partly to blame for the disastrous performances of their companies.

Particularly at Baldwin and Lima, managers retained their perception of the diesel as adjunct technology, an attitude fostered by their expertise in steam locomotive production and their success in fulfilling the varied steam-locomotive design requirements of railroad operating officials. This, in turn, made it far more difficult for managers to design the research and development programs, plant rationalizations, and marketing strategies necessary to make successful their forays into the diesel locomotive market. Lima and Fairbanks-Morse lacked the financial resources necessary to ensure continued success in the diesel locomotive industry. And, by the end of World War II, Baldwin was light years behind EMD in terms of locomotive technology and manufacturing methods and thus could never hope to catch up.

A great deal of recent historical debate has centered on the roles and relative importance of large-scale mass-production firms and customized small-batch producers. Scholars such as Alfred D. Chandler Jr. have asserted, correctly, that giant firms, with their standardized production and well-defined organizational routines, have contributed greatly to the strength of the American economy. Others, such as Michael Piore and Charles Sabel, have been sharply critical of the large American corporation, particularly in light of the recent poor economic performance of many such organizations, and they have instead emphasized the continuing importance of smaller, more adaptable firms. Philip Scranton has also emphasized the strength of small-batch producers and has attributed Baldwin's success in the steam locomotive industry to that company's custom manufacturing capabilities.[142]

While some industries have experienced a recent transition from mass production to custom production, the situation in the locomotive industry was just the reverse.[143] Small-batch custom manufacturing constituted the most effective production strategy within the context of the steam locomotive industry. In this particular case, radical technological change mandated radical changes in production practices, ensuring that standardized near-mass-production became the most viable strategy within the context of the diesel locomotive industry. As such, ALCo, Baldwin, and Lima not only had to change their products and their production facilities, they had to embrace an entirely new manufacturing philosophy. The failure of all three of these companies in the diesel locomotive industry indicates the difficulties involved in the transition between customized manufacturing and standardized mass production.

Ultimately, particularly at Baldwin and Lima, executives made a crucial mistake by believing that the strength of their companies lay in locomotive production, and that it did not matter whether these were steam or diesel locomotives. Instead, the real strength of their companies lay in the ability to produce small batches of custom-manufactured capital goods. Steam locomotives matched these production capabilities; diesels did not.

VII

The Era of Oligopoly

WITH THE CESSATION of diesel locomotive production at Baldwin, Lima, and Fairbanks-Morse, the industry assumed its current duopolistic structure. Duopoly emerged in the locomotive industry for several reasons. The small "feast-or-famine" market for locomotives, combined with the capital-intensive nature of locomotive production, ensured that a single large producer could satisfy railroad demand. A lone manufacturer would certainly run afoul of the U.S. Justice Department, a situation EMD experienced during the mid-1960s, and this precluded the existence of monopoly. Since railroads were hardly technological innovators in their own right during the postwar era, their executives believed that technological progress in the locomotive industry demanded competition. As a result, railroads frequently split their locomotive orders between EMD and Alco, even though they realized that the former company produced better locomotives, and even though this policy reduced standardization within railroad motive-power fleets. Railroads had little interest in making Alco an equal competitor to EMD. Instead, they merely wanted another company in the industry to keep EMD competitive and technologically progressive. As Alco discovered in the 1960s, American railroads did not care which company occupied this secondary position. During that decade, railroads forsook Alco locomotives for the technologically superior products offered by General Electric, a longtime participant in the locomotive industry, but, as of 1960, a new entrant into the large mainline freight diesel locomotive market. As a result, Alco experienced a rapid decline in market share during the 1960s and ended locomotive production in 1969. Since then, the American diesel locomotive industry has remained a duopoly consisting of GE and GM-EMD, with both companies vying for the top position in the locomotive market.[1]

EMD Dominates the Diesel Locomotive Market

During the 1960s, EMD improved its diesel locomotive technology, enabling the division to retain the largest share of a shrinking diesel-locomotive market. Alco posed even less of a threat than it had during the 1950s;

and GE, despite its early success, posed no immediate danger to EMD's market share. The real threat came from the federal government, which attributed EMD's 80-plus percent market share to monopoly power rather than the successful integration of production, marketing, and managerial skills. The Justice Department posed a serious threat to EMD's market dominance, as well as to its long-standing ties to General Motors.

While the American economy as a whole was sluggish during the late 1950s and early 1960s, the railroad industry performed especially poorly. Improved highways and inland waterways, combined with the speed and flexibility of long-haul trucking services, reduced the railroad's share of freight traffic. The increasing popularity of air and auto travel resulted in massive losses in railroad passenger traffic. By 1961, railroad earnings had fallen to one of their lowest levels ever. Capital expenditures for railroad equipment shrank from just over $1 billion in 1957 to $427 million in 1961. While American railroads placed in service 1,015 new locomotives in 1957, they purchased only 288 in 1961. This declining demand forced EMD, Alco, and, later, GE to be particularly innovative in the development of new locomotive designs.[2]

EMD improved both its diesel engines and its locomotives in response to the tightening of the locomotive market. In 1961, EMD introduced the 2,250-hp GP-30 and secured orders for more than a hundred of these units within the first month of its availability. The even more popular 2,500-hp GP-35 followed the GP-30 two years later. EMD equipped these units with a more powerful variant of the Model 567 engine and intended them to serve in both branch line and mainline service.

By the early 1960s, however, EMD engineers had concluded that the Model 567 engine, in production since 1938, was no longer adequate for the increasing horsepower requirements of the railroads. Accordingly, in June 1965 EMD introduced the Model 645 engine, with a cylinder size that had been increased from 567 to 645 cubic inches. Designers took pains to ensure that many components of the new Model 645 engine were interchangeable with those of the older Model 567. Along with the larger engine, EMD introduced a completely new locomotive line, consisting of nine models ranging from 1,000 hp to 6,000 hp. EMD predicted that these new locomotives would pay for themselves in one year, based on savings in repair costs alone, and boasted that three of these new locomotives could replace seven older units.[3]

Taken together, the development of the Model 645 engine and the creation of the new locomotive line constituted the longest research and development program and the largest capital investment expenditure in EMD's history and laid the foundation for an additional twenty years of market dominance. EMD timed well the changeover from the Model 567 to the

Model 645. By 1965, railroad earnings were beginning to improve, and railroads were anxious to take advantage of the cost savings offered by EMD's new diesel line. In addition, diesels purchased from EMD and other builders in the postwar dieselization boom were rapidly approaching retirement age, and the limited interchangeability between the two engine models was thus not a serious issue. In contrast, ALCo's switch from the Model 244 to the Model 251 engine in 1953, at the end of the dieselization rush, showed unfortunate timing. Railroads that had just spent millions of dollars on ALCo locomotives discovered that these new units were almost immediately obsolete.[4]

The Federal Government Intervenes

The federal government threatened EMD's market dominance during the early 1960s, however, and the Justice Department's efforts to prosecute General Motors for alleged antitrust violations could have seriously weakened EMD's competitive position in the diesel locomotive industry.[5] While there is some evidence to support the government's accusations, it is clear that the Justice Department was incorrect in its assumption that EMD's large market share constituted a monopoly or that EMD had obtained that market share through illegal means. The federal government exhibited a fundamental ignorance about the economic realities of the locomotive industry and confused first-mover advantages with monopoly. Given the impressive advances that had been made in the field of economics since the passage of the Sherman and Clayton antitrust acts, it is difficult to understand why the government decided to prosecute EMD.[6]

During the years after World War II, the federal government showed increasing interest and concern regarding the size and enormous economic power of General Motors and its various subsidiaries. As early as 1955, the Subcommittee on Antitrust and Monopoly of the Senate Judiciary Committee investigated GM's corporate structure and business practices and issued a report entitled *A Study of the Antitrust Laws: Bigness and Concentration of Economic Power—A Case Study of General Motors Corporation*. The report concluded that GM's "size was of paramount importance. . . ." to EMD's market dominance.[7]

A series of legal actions bore out the government's aversion to GM's size. In 1956, the Justice Department filed an antitrust suit against GM, charging that its 89 percent share of the intercity bus industry constituted a violation of the Sherman Act. In June 1957, the Supreme Court declared illegal DuPont's ownership of 23 percent of GM. Two years later, in October 1959, the government claimed that GM's involvement in the construction equipment industry, through its Euclid Division, violated the Clayton Act. Industry experts concluded that these developments were part of a larger effort to attack the periphery of General Motors, rather than confront the company directly.[8] For example, in 1961 *The Journal of Commerce* stated, "The apparent attempt of the Justice Department to divest General Motors of its Electro-Motive Division bears all the ear-marks of an attack on 'bigness as such,' rather than a bona fide move to check discrimination or other harmful or illegal practices...."[9]

On April 12, 1961, EMD became a part of this "attack on bigness" when a federal grand jury indicted GM for alleged violations of Section 2 of the Sherman Act regarding its activities in the locomotive industry. In February 1959, Attorney General William P. Rogers, an Eisenhower appointee, initiated an investigation against EMD, and this led directly to the 1961 indictment. A grand jury, empaneled on November 17 of that year, met for the next seventeen months, while the new Kennedy administration continued and expanded this antitrust action.[10]

Specifically, the indictment charged that EMD had sold locomotives at a loss and that it had adjusted its locomotive prices to undermine its competition; that EMD had produced locomotives at a rate that no other company could match; that EMD's competitors could not equal GM's financial resources, level of investment, advertising expenditures, or ability to establish service and repair facilities; and that GM could obtain better terms for parts and supplies than could any of its competitors. According to the Justice Department, EMD also offered unfair financing and made attempts to monopolize the diesel locomotive rebuilding market. Finally, the Justice Department charged GM with reciprocity, using its position as the largest rail shipper in the United States to threaten railroads that refused to purchase EMD locomotives and reward those that did.[11]

While this criminal suit wound its way through the legal system, the federal government, on November 14, 1963, filed a civil antitrust suit against General Motors, charging that company with violating the Sherman and Clayton Acts. In addition to the now-familiar reciprocity charge, the government claimed that GM's 1930 purchases of the Winton Engine Company and the Electro-Motive Company violated the Clayton Act's restrictions on acquisitions that tended to create monopoly. The Justice Department also reiterated its belief that EMD offered unfair financing terms and that it sold locomotives at a loss.[12]

Although GM officially denied all of these charges, there is some evidence that EMD did indeed engage in questionable business practices, particularly regarding reciprocity. As early as 1944, ALCo president William Dickerman suggested that GM had used reciprocity to compete unfairly against

THE ERA OF OLIGOPOLY 131

his company, but he was unable to furnish any proof. Railroad correspondence files offer more telling indictments. When the Pennsylvania Railroad purchased an EMC switcher in 1937, the company made certain to inform the traffic directors of both Chevrolet and Fisher Body.[13] In 1939, when the president of the Louisville and Nashville considered purchasing EMC locomotives, he

> told the sales representative of the Electro-Motive Corporation that I hoped they would consider this purchase in such a way as to influence a larger amount of competitive traffic via this railroad. He said the Electro-Motive Corporation itself had very little [traffic] that they could influence via this railroad. He further said, however, that all of their sales were brought to the attention of the General Motors Corporation. I am hoping, therefore, that you can use this purchase advantageously in the matter of securing traffic.[14]

The reassurances of this EMC salesman probably did not represent official GM policy. In addition, this incident occurred during EMC's formative years, before GM incorporated it as a division in 1941. GM's increasing control over its subsidiary, a process described in chapter 5, was in part designed to prevent potentially embarrassing situations of this nature. Furthermore, the Justice Department, despite its lengthy investigation, never found this information, which, in any case, would not have been enough to convict GM of any significant violation of antitrust law.[15]

After years of effort, the Justice Department amassed little concrete evidence against GM. In addition, Justice Department prosecutors made a number of mistakes during the course of their investigation. They filed the initial 1961 charges in New York, both because that city was close to Washington and because they felt that the New York court had a stronger antimonopoly stance than its counterpart in Chicago. An exasperated New York District Court judge almost immediately ordered the case transferred to Chicago, where La Grange was located, adding that "monopoly may have been thrust upon [EMD]."[16] In his opening arguments before the court, Department of Justice Chief Prosecutor Sanford M. Litvack frequently referred to Alco as "Alcoa," a company that produced aluminum, not locomotives. Based on a lack of evidence and a growing realization that EMD's success was a result of investments in integrated production, distribution, and management rather than monopoly, the Justice Department dropped all charges against GM and EMD in December 1964.[17]

Two factors had considerable impact on the Justice Department's decision to drop the antitrust suits against EMD. First, scores of railroad industry executives, who understood that EMD offered both a better product and better service than any of its competitors, testified that EMD's success had been based on these factors alone, rather than economic pressure by GM.

According to GM's legal counsel, this railroad testimony "had a subtle and persuasive influence which undoubtedly affected the Government's decision to dismiss the case...."[18] Second, GE's rapid success in the diesel-locomotive industry in the early 1960s offered clear evidence that EMD had not created insurmountable barriers to entry.

GE Enters the Large Diesel Locomotive Industry

The decade of the 1960s witnessed the entrance and subsequent rapid success of a new competitor in the large-diesel locomotive industry. General Electric used its decades of experience in the small-diesel and export-diesel markets to great advantage in its efforts to displace Alco, its former production partner, from the industry. GE initially explored the possibilities of the diesel locomotive market because managers in the Locomotive and Car Equipment Department realized that serious production and managerial difficulties existed at Alco. In response, GE first terminated the long-standing joint-production agreement, largely to protect itself from the embarrassment and expense associated with Alco's mistakes. In evaluating Alco's difficulties, GE executives understood that their company could earn substantial profits in the locomotive industry by making their own locomotives in competition against both Alco and EMD. GE's ability to produce better locomotives than Alco, along with its managerial and marketing expertise, allowed that company to drive Alco from the diesel-locomotive industry in less than a decade.

As discussed in chapter 5, GE used World War II as an opportunity to implement standardized production techniques for its small diesel switchers, and the company soon reaped the rewards. In 1946, for example, GE secured orders for 178 diesel locomotives, all 600 hp or less. Forty-eight of these were for service on American railroads, while an additional 104 were destined for domestic industrial use. In 1954, GE offered forty different locomotive designs, more than ever before in its history. Two years later, GE introduced a new line of "Universal" export locomotives, ranging from 400 hp to 1,980 hp. The largest of these would have qualified as road freight locomotives, had they been sold in the domestic market. One model in particular, the 6,000-hp, four-unit XP 24–1, first tested in September 1954, later served as the progenitor of GE's domestic road freight diesel locomotive line.[19]

Increased locomotive sales encouraged GE to make gradual improvements to its switchers in the postwar period. In 1952, GE expanded its production capabilities by constructing a new traction motor research and development facility. A year later, as part of a larger corporate effort to expand into the industrial automation sector, GE consolidated its traction motor and generator production at Erie. By 1955, GE was marketing its own electrical equipment independently of Alco.[20]

GE also expanded its support services. In 1949, GE equipped a facility in Richmond, Virginia, to repair diesel locomotive traction motors and generators. Five years later, GE established a complete locomotive rebuilding facility at Erie. By 1957, the company maintained locomotive parts distribution centers in Chicago, Atlanta, St. Louis, and Los Angeles. GE also operated major locomotive overhaul facilities in Atlanta; Boston; Chicago; Cleveland; Dallas; Minneapolis; New York City; Pittsburgh; Portland, Oregon; Salt Lake City; and San Francisco. In addition, railroad customers could obtain many parts and services, twenty-four hours a day, at thirty-seven other GE facilities in the United States—a support network that far exceeded what Alco was able to offer. Initially, GE designed these facilities for its small diesel switchers but could, and later did, convert them to service facilities for large diesel locomotives. By 1971, this service network had grown to include five regional parts centers (at Erie, Atlanta, Los Angeles, Minneapolis, and St. Louis) along with sixty-seven Service Shops.[21]

GE matched these physical plant investments with improvements in marketing services. Like the other builders, GE offered a locomotive school. In addition, GE, rather than Alco, conducted marketing studies to determine what type of GE or Alco-GE locomotives best suited a particular railroad's operating requirements. GE employees also advised railroads on the most efficient design for new diesel locomotive service and repair facilities. As was the case with GE's service and repair capabilities, these marketing efforts outweighed Alco's marketing capabilities and ensured that GE had ample practice in the development of its own organizational strengths prior to the company's introduction of large freight diesels in 1960.[22]

"Damaging Their Reputation, and Ours"

GE did not base its decision to enter the large diesel locomotive market solely on its expanded manufacturing and marketing capabilities, however. Alco provided the final impetus, since its manufacturing ineptitude provided a continual source of embarrassment for GE—and also offered an opportunity too good to pass up. During the postwar period, railroads made numerous complaints about ALCo-GE locomotives.[23] Since all ALCo-GE locomotives displayed a prominent GE emblem on their sides, these problems, although rarely caused by GE, naturally concerned officials of that company. J. M. Symes, the Pennsylvania Railroad's vice president of operations, made a point of informing a GE executive "that the Alco diesels are much more expensive to maintain than the EMD diesels and that we cannot afford to buy Alco diesels in view of that situation."[24] A 1954 GE marketing report concluded, "Present trends indicate that it is imperative to get control of the complaint situation if we are to stay in business. The continued downward trend in business volume is traceable in most cases to poor product performance."[25] A year later, another GE report noted how the "stigma of lack of reliability earned by the Alco passenger units on the Santa Fe is still following Alco and damaging their reputation, and ours."[26]

Since it was virtually powerless to remedy the chronic managerial and manufacturing difficulties at Alco, GE did its best to distance itself from that company. In 1953, GE terminated the joint-production agreement, although the company continued to supply electrical equipment for Alco diesels. In addition to removing the GE nameplate from inferior Alco diesels, this decision allowed GE to enter the large diesel market in its own right.

Unlike the haphazard entrance of ALCo, Baldwin, Lima-Hamilton, and Fairbanks-Morse into the diesel locomotive industry, GE carefully planned its introductory campaign and committed itself only when its managers were certain that their company could capture a substantial market share. In the mid-1950s, GE surveyed thirty-three railroads to determine their future motive-power needs. From these studies, GE realized that railroads wanted reduced maintenance costs above all other qualities. Other considerations, in decreasing order of importance, included reduced operating costs, increased reliability, greater pulling power, lower cost per horsepower, and increased horsepower per pound of locomotive weight.

Based on these marketing studies, GE concluded that it could participate successfully in the locomotive industry if it could reduce operating costs by at least 35 percent and lower locomotive costs from $100 per horsepower to $80 per horsepower or less. Significantly, potential problems in the design and construction of these new locomotives caused little concern among GE engineers; they assumed, correctly, that they would have no difficulty designing a locomotive that could meet these requirements. Based on these marketing forecasts, GE concluded that "the market for road diesel-electric locomotives, therefore, becomes a *very inspiring* picture . . ." and predicted annual sales ranging from $160 to $200 million.[27] This would ensure GE at least a 16 percent return on capital. The company estimated that significant sales of second-generation replacement diesel locomotives would begin in 1960, reach a peak in 1965, and remain at a high level for the following ten years.[28]

As a result of these substantial potential profits, GE executives committed their company to the large diesel freight locomotive market. GE had initially planned to offer the XP 24-1, first tested in September 1954, as an export locomotive, but the company reassigned this demonstrator as a secret test unit, once GE had decided to enter the domestic road diesel market. This locomotive consisted of four connected units operating under unified control. A 1,500-hp Cooper-Bessemer diesel engine powered each unit, giving

THE ERA OF OLIGOPOLY   135

the entire demonstrator an impressive 6,000 hp. GE supplied the electrical equipment. In April 1960, GE introduced the regular production version of the XP 24-1. The U-25B was essentially a single-unit version of the XP 24-1, though it lacked a streamlined carbody and boasted a more powerful 2,500-hp, sixteen-cylinder, four-cycle Cooper-Bessemer diesel engine. GE's use of the XP 24-1 as a covert test bed had been so successful that the new U-25B "caught the railroad industry by surprise. Most railroad men thought the test locomotive was only a research lab for service testing of components."[29] Even though outside contractors supplied critical locomotive components, GE assumed all responsibility for the locomotive, including that of marketing the finished product. The U-25B was an immediate success, and GE received orders for more than two hundred of these locomotives, worth some $31 million, before the end of 1960.[30]

The success of the four-axle U-25B prompted GE to integrate production and offer additional locomotive models. The six-axle U-25C debuted in August 1963, followed by three additional models in 1966. By the end of the decade, GE offered a complete locomotive line. GE's market share increased from 2.8 percent in 1960 to 18.7 percent in 1967, while Alco's market share declined from 13.1 percent to 6.5 percent in the same period (some of GE's gain came at EMD's expense). The success of the U-Series locomotives prompted GE to begin manufacturing its own diesel engines in 1964, thus giving the company the distinction of being the second integrated manufacturer of diesel locomotives in the United States. A year later, GE elevated its Locomotive and Car Equipment Department to divisional status.[31]

GE's rapid success in the locomotive industry occurred because the company's locomotives were far better than Alco's products, though considerably inferior to those offered by EMD. Tests on the Louisville and Nashville Railroad, conducted in May 1960, indicated that GE locomotives had more power and better acceleration than their Alco counterparts. Robert N. Cotton, who was the first black locomotive engineer on the L&N and one of the first in the United States, complained that Alco locomotives were difficult to operate and thought that they were "made [to] aggravate the locomotive engineer." He also observed that GE locomotives were "the most sophisticated" of those that he had operated.[32]

By the early 1970s, however, railroads had discovered serious quality control and maintenance difficulties in early GE locomotives. While these locomotives were not as yet technologically competitive with their EMD rivals, they still outperformed Alco diesels, and that alone was enough to give GE the secondary market position in the locomotive industry.

General Electric entered the large diesel locomotive market in a direct response to Alco's manufacturing difficulties, inadequate technology, and poor quality control. During the late 1950s and into the 1960s, GE coveted EMD's huge market share but, for the time being, Alco was the easier target.

In wresting market share from Alco (and, to a lesser extent, from EMD), GE utilized decades of experience in the industrial diesel and export diesel markets. The company explored carefully both railroad needs and its own capabilities in order to determine whether or not it should enter the large diesel locomotive market. Once the company had decided on this course of action, GE developed locomotives that were superior to those offered by Alco. GE thoroughly tested a prototype unit in what amounted to total secrecy, an impressive accomplishment considering that the object in question was two hundred feet long and weighed five hundred tons. By using Cooper-Bessemer engines, GE reduced its initial investment yet retained control over the final product. Then, when its locomotives became commercially successful, GE integrated engine production with its existing locomotive manufacturing capabilities, thus ensuring control over production and profits. GE used its organizational capabilities to replace Alco as the secondary producer in the duopolistic locomotive industry within three years and forced Alco to end production in nine.

EMD and GE Contest Market Dominance

Although GE easily displaced Alco from the diesel locomotive industry during the 1960s, EMD proved a much more formidable competitor. And, while GE diesels clearly outperformed their Alco counterparts, they were initially no match for EMD's products and indeed quickly acquired a reputation for poor quality control, low reliability, and high maintenance expenditures. GE worked to overcome these problems, and by the 1980s, its success rivaled EMD's. At the same time, both builders suffered from wild fluctuations in locomotive demand that caused General Motors to consider terminating its involvement in the diesel locomotive industry.

A variety of technical problems beset the GE U-Series locomotives. These included poor wiring, fuel line ruptures, malfunctioning blower fans, and overheating diesel engines. While EMD locomotives were available for revenue service 97 percent of the time, those produced by GE managed only an 88 percent level of availability. GE's U-Series locomotives averaged an abysmal fifty days between repairs.[33] Railroad executives complained that these "GE locomotives have been so unreliable that only a mother could love them."[34] Even J. Robert Malone, a GE locomotive marketing executive, admitted that his company's products were "not that good."[35]

Wildly fluctuating, but generally declining, locomotive demand exacerbated GE's difficulties and caused problems for EMD as well. Since the 1930s, locomotive builders had succeeded admirably in their efforts to increase locomotive horsepower to higher and higher levels. Paradoxically,

THE ERA OF OLIGOPOLY   137

this meant that railroads could haul the same amount of freight with fewer locomotives. As locomotives increased in size, complexity, power, and price ($1.5 million per unit by the early 1990s, compared to approximately $250,000 in 1945), locomotive orders simultaneously declined. Economic recessions, changes in the regulatory climate, railroad mergers, and periodic slumps in rail traffic could cause locomotive orders to plunge even more precipitously. The 1973–74 energy crisis curtailed locomotive orders so severely that railroads ordered fewer than five hundred units in 1976. Orders then rose to approximately seventeen hundred locomotives in 1979, then fell to barely two hundred units in 1983. Total U.S. locomotive production for both domestic and foreign markets equaled only 131 units in 1987. Given such fluctuations, product innovation and customer responsiveness became even more vital to success in the locomotive industry but did not themselves guarantee that GE—or even the once-mighty EMD— would prosper.[36]

During the early 1970s, American railroads demanded locomotives that combined reliability with easier maintenance, and both builders responded to these requests. EMD introduced its "Dash-2" series locomotives in January 1972, while GE launched a new locomotive line in July 1976. These locomotives reduced maintenance costs but, in the case of the GE models, improved reliability only marginally.[37]

As the decade of the 1970s wore on, rising fuel costs and increasing public and governmental concern regarding pollution caused many railroads to shift their locomotive design priorities. While average annual fuel costs per locomotive had been $15,000 in 1969, they rose to $97,000 ten years later. As such, customers increasingly demanded fuel-efficient, low-emission units. GE benefited from this change in priorities, since its four-cycle diesel engine was inherently more fuel-efficient than EMD's two-cycle version. Since builders typically employed increasingly sophisticated electronic controls to reduce fuel consumption and emissions, GE's familiarity with electronics and electrical equipment technology gave that company's locomotives a further edge over EMD.[38]

GE took the lead in the production of high-horsepower fuel-efficient locomotives. During the early 1980s, GE reduced its locomotive line from fourteen models to nine and halved its available "product options" (i.e., the arrangement of optional features) to 277. The company spent $316 million on improvements to its Erie manufacturing facilities, installing robots and computerized quality-control equipment and creating a Learning and Communications Center to facilitate product development and marketing. Most important, GE introduced a new "Dash-8" locomotive line in January 1983. These locomotives employed sophisticated microprocessor technology to regulate optimal engine speed, inject precise amounts of fuel into the cylinders, control wheel slippage, and diagnose malfunctions. The Dash-8 locomotives quadrupled the onboard electronics found on a standard diesel locomotive and left EMD struggling to catch up. In 1983, EMD managed only a 40 percent share of the locomotive market, losing market dominance for the first time since 1935. GE's decision, in January 1989, to replace the head of its locomotive operations led to further production efficiencies and quality-control improvements.[39]

While GE was improving the sophistication and reliability of its locomotives, EMD was experiencing severe production and quality control difficulties. EMD's SD-50 locomotives were extremely unreliable, for example, and many had to be extensively rewired at EMD's expense.[40] In order to win back customers from GE, EMD spent $60 million between 1980 and 1984 to develop the Model 710 diesel engine as a replacement for the outdated Model 645. Tooling costs for the new engine consumed another $78 million. EMD installed these engines in its new "60-Series" locomotives, which employed onboard microprocessor controls to increase fuel efficiency. EMD also recognized that decades of unrivaled success had caused its own operational routines to ossify. Accordingly, EMD sought suggestions from railroads, pledged to increase its responsiveness to its customers, and eliminated many traditional job classifications and work rules. Peter Hoglund, a GM vice president and EMD's general manager, hoped that "the new organizations will erode the old, traditional, sometimes arrogant attitude that we have developed over the years." EMD's new product and managerial innovations enabled the division to reacquire market dominance in 1984, only to be surpassed again by GE in 1987.[41]

New electrical equipment technology, in the form of alternating-current traction motors, strengthened EMD's position in the locomotive industry. Before the late 1980s, virtually all diesel locomotives built in the United States employed direct-current traction motors. By the 1980s, however, DC traction motors had become a reverse salient in locomotive design; they were too large, too complicated, and too unreliable to cope with more powerful and more fuel-efficient locomotive designs. European railroads had already made extensive use of AC traction motors. Despite their higher initial cost, these motors offered greater reliability, easier maintenance, better traction, and a lower weight-to-horsepower ratio. GM modified European AC technology, obtained from Siemens AG, to suit American operating practice. While American freight railroads exhibited a keen interest in AC traction motors, none wished to gamble millions of dollars on this new technology. Instead, GM sold its first AC-motored F69PH locomotives to passenger carrier Amtrak in 1989. The success of these locomotives led, in March 1993, to the first large order for AC-equipped EMD freight locomotives. GE introduced its own AC locomotives, but its technological expertise in this area still lagged behind EMD's.[42]

THE ERA OF OLIGOPOLY

Despite the capital-intensive nature of diesel locomotive production and despite the feast-or-famine tendencies of the locomotive market, new companies contemplated penetration of the locomotive duopoly during the 1980s and 1990s. Perhaps the most far-fetched of these proposals involved the American Coal Enterprises (ACE)-3000, a design for a modern, fuel-efficient, low-maintenance, coal-fired steam locomotive. While promoters of the ACE-3000 understood that the United States possessed vast untapped coal reserves, they underestimated the steam engine's inherent thermal inefficiency and, more pragmatically, lacked the $30–$45 million necessary to develop a working prototype.[43]

The Morrison-Knudsen Corporation presented a more viable threat to GE and EMD. Its subsidiary, M-K Rail, had achieved considerable success in the railroad equipment, freight car, mass-transit, and locomotive rebuilding markets. By 1990, this company had rebuilt or repowered more than nine hundred locomotives from various builders. Because EMD and GE solicited large orders for high-horsepower locomotives from major railroads, they often neglected smaller carriers, and M-K Rail attempted to take advantage of this market. The company was also anxious to sell locomotives in Latin America and the Pacific Rim. In 1992 M-K Rail signed an agreement with Caterpillar to provide engines for both diesel and natural-gas powered locomotives. The company completed its first new locomotives in 1994 but found few customers. By the end of that year, M-K Rail had laid off 5 percent of its employees, and three stockholders had filed suit against Morrison-Knudsen, accusing the parent company of "misleading them about the viability of the rail business." Morrison-Knudsen clearly underestimated the size of the barriers to entry that existed in the locomotive industry and suffered accordingly.[44]

While Morrison-Knudsen failed to penetrate the diesel locomotive industry, General Motors contemplated exiting that industry. Although EMD led GE in AC technology, the division's wage rates, manufacturing costs, and prices were higher than GE's, and by 1993, GE diesels were outselling EMD products two-to-one. Declining locomotive orders caused EMD to shut down La Grange in 1983, in 1987, and again in 1988, temporarily shifting orders to GM Diesel in London, Ontario. By 1991, more than one million square feet of space sat unused at La Grange—a sight eerily reminiscent of Baldwin's grossly oversized Eddystone facility. Faced with this situation, EMD desperately expanded into locomotive inspection, maintenance, and rebuilding services, along with locomotive leasing. EMD also allowed its customers greater control over the design and manufacturing process. In 1994, EMD furnished the Consolidated Rail Corporation (Conrail) with diesel locomotive "kits," which the railroad, not the locomotive builder, assembled in its Altoona, Pennsylvania, shops.[45] EMD offered many additional options on its locomotive models and became increasingly willing to work with railroad motive power officials to modify existing designs substantially, a process that faintly echoed the glory days of the steam locomotive industry.[46]

Sluggish locomotive demand coincided with an effort by GM to concentrate on its core automobile and light truck markets. GM spun off its heavy truck business as a joint venture in 1986. Detroit Diesel followed two years later. In 1991, GM sought to dispose of EMD either as a joint venture or through outright sale. Possible partners or purchasers included ASEA Brown Boveri and Caterpillar. GM was never terribly aggressive in its attempts to shed EMD, and no outside company expressed serious interest in the diesel locomotive component of GM's operations. As of this writing, General Electric and the Electro-Motive Division of General Motors still control the diesel locomotive industry in the United States.[47]

Alco's Decline

By the early 1960s, the newly created Alco Products Company was in serious financial trouble. Total sales, which had been a respectable $440 million in 1953, had plummeted to $89 million in 1961. Total earnings fell 92 percent between 1955 and 1961. The wide variety of custom-engineered and custom-built products, which were intended to equal locomotive sales, instead comprised only 20 percent of revenues and an even smaller percentage of profit. In addition, Alco executives unwisely diversified into customized product lines, such as nuclear power, that required extensive capital investments and immense technical expertise—assets that were in short supply at Alco. Alco's locomotive division performed little better. Employment at the Schenectady plant declined from ten thousand in 1951, the peak year of ALCo's diesel production, to two thousand in 1960. Although not in imminent danger of bankruptcy, Alco nevertheless required a substantial change in corporate strategy.[48]

On September 1, 1962, Alco announced its second reorganization in seven years. The company instituted a line-and-staff managerial structure, and a multidivisional corporate structure replaced the earlier centralized, functionally departmentalized one. Although the business historian Alfred D. Chandler asserts in *Strategy and Structure* that most well-managed companies allow the structure of the company to follow its strategy, Alco adopted the opposite approach.[49] The reorganization created a multidivisional structure in anticipation of an expansion into new product lines. Although Alco acquired no new product lines, the company did divest itself of many of its unprofitable subsidiaries.[50] During 1962 and 1963, Alco sold all of its custom-engineered product lines, including heat exchangers, nuclear-power components, and oil field products. Alco continued to manufac-

THE ERA OF OLIGOPOLY   141

ture standard-design machinery, including locomotives, diesel engines, forgings, and springs.⁵¹

This concentration on locomotive production and the concomitant disposal of unrelated product lines seemed to defy logic, especially given the stagnation of locomotive demand at the time of Alco's reorganization. Three factors supported Alco's decision, however.

First, since EMD was generally too busy filling domestic orders to give much consideration to foreign demand, Alco attempted to exploit this market. Alco adopted a dual strategy for the penetration of foreign markets. In countries with large rail networks, Alco licensed firms to manufacture locomotives locally. Typically, these firms already produced similar items, such as diesel engines or railway equipment. In 1959, Alco signed an agreement with the London-based Associated Electrical Industries, Ltd. "for joint collaboration in the design, manufacture and sale of diesel-electric locomotives [of more than 900 hp] for world markets."⁵² Alco opened a London sales office in 1960 and in the same year signed an agreement with Davey, Paxman and Company, Ltd., for the joint production of diesel engines in Great Britain—too late to give Alco an effective position in the British market. In 1964, Alco issued a license to the Societe des Forges st Chantiers de la Mediterranee for the production of its diesel engines in France. Alco maintained a similar agreement with Compania Euskalduna de Construccion y Reparacion de Buques in Spain. A. E. Goodwin, Ltd., built diesel locomotives to Alco designs for the Australian market. In February 1962, Alco executives signed a $200-million agreement with the government of India for the manufacture of Alco-designed locomotives at a plant in Varanasi, Upper Pradesh. By the mid-1960s, half of all Alco locomotives destined for foreign markets were built abroad.⁵³

Nations with low rail mileage and small locomotive demand had their diesels built in Alco's Schenectady facility. Alco executives clearly favored U.S. production over the issuance of production licenses to foreign firms. According to Board Chairman William G. Miller, "Direct sales of equipment, whereby locomotives are built in the United States and shipped overseas, are uppermost in our minds when dealing abroad. However, we have fully faced up to the facts of life, and have adapted our operations to meet the changing requirements of those nations that increasingly are seeking ways and means of self-manufacture."⁵⁴

Alco's penetration of the international market was only modestly successful. The granting of production licenses to foreign manufacturers, while profitable, did little to increase output at Schenectady. Foreign orders tended to be smaller than those placed by American railroads. In addition, foreign railroads often exhibited wide variations in track gauges and clearance restrictions, factors that made standardized production difficult. Moreover, by the late 1950s, as its domestic orders dwindled, EMD concentrated its resources on the foreign market, to the detriment of Alco. Thus, in 1959, EMD surpassed Alco in sales of diesel locomotives for export. In all years prior to 1959 EMD averaged 27 percent of the foreign market, while in 1960 it had a 68 percent share of all foreign orders. Finally, Alco ran headlong into the locomotive development policy practiced by its domestic production partner, General Electric. In a policy opposite to that practiced by EMD, GE used foreign railroads as test markets for the development of large, high-horsepower locomotives, thus gaining experience that it later used to push Alco out of the secondary domestic market position.[55]

The second justification for Alco's decision to focus on locomotive production arose from hopes that the federal government might reduce EMD's market share. This was not to be, however, since the Justice Department's 1961 and 1963 antitrust suits did not force General Motors out of the locomotive business. In December 1964, the government dropped both antitrust suits against GM, thus ensuring that EMD was not legally obligated to alter its competitive behavior. However, the threat of future prosecution may have encouraged EMD to allow its competitors to increase their market shares slightly. In addition, GM executives understood that voluntary efforts to restrict EMD to no more than 75 percent of the market would better enable the division to cope with demand fluctuations. Alco could not take advantage of this opportunity, however, because the market share given up by EMD was soon appropriated by General Electric.[56]

Finally, and most importantly, Alco executives predicted that domestic locomotive demand would begin to increase over the next few years. Their predictions were correct. Locomotives purchased between 1945 and 1955 were nearing the end of their useful lives, since diesels typically had a service life of approximately twenty years and were fully depreciated after fourteen. Technological improvements, which boosted horsepower and lowered maintenance costs, persuaded many railroads to replace their aging diesel fleets en masse. Thanks in part to new locomotive depreciation schedules issued by the Treasury Department in July 1962, locomotive demand surged forward. The years 1965 and 1966 constituted two of the best years in the history of the industry, with more than 1,200 new and rebuilt units added to railroad motive power fleets each year.[57]

Alco exploited this increased demand with its Century Series locomotives. Alco executives assigned the development of these models to the company's newly created Advanced Locomotive Design Group, which was kept separate from the regular Production Engineering Department. The company first offered these models in January 1963, claiming that they lowered operating costs up to 40 percent and maintenance costs up to 50 percent, compared to locomotives that had been in service ten years or more.[58] Still, these locomotives could not compete against their EMD and GE counterparts. On the Northern Pacific, Alco units cost nearly eleven cents per loco-

THE ERA OF OLIGOPOLY 143

motive mile to maintain, while EMD locomotives cost just over six cents per mile. The railroad also concluded that GE locomotives contained better electrical equipment and less complicated electrical controls than Alco products. Continuing deficiencies in Alco's service network compounded these problems.[59] The Northern Pacific's chief mechanical officer complained that "the repair parts situation for Alco has never been satisfactory and apparently there is no improvement in sight.... Alco has the poorest Service Organization of all the Locomotive Builders."[60]

Alco was not the only firm to seize the market opportunities created by this sudden increase in locomotive demand. In June 1965, EMD introduced nine entirely new locomotive models, a number that increased to fourteen by the end of the year. These locomotives incorporated the larger and more powerful Model 645 engine as a replacement for the Model 567. EMD offered a massive twenty-cylinder version of this engine, a product that Alco did not match.[61]

However, it was General Electric, not EMD, that turned the 1960s into a bitter decade for Alco. When GE entered a market that could be served adequately by two large producers, its superior technology and greater production efficiency forced Alco out of diesel locomotive production by the end of the decade. This situation, in which Alco competed directly with its only supplier of electric equipment, was unusual in American business, though not unique.

Following the 1962 reorganization, Alco executives intended to diversify their operations by creating a conglomerate, a form of corporate organization then very much in vogue.[62] Although they possessed both the cash and the necessary multidivisional corporate structure, Alco executives did not purchase any additional companies or product lines.

Instead, Alco itself became part of a larger firm. In July 1964, the Worthington Corporation, which had purchased Alco's feedwater heater business less than two years earlier, offered to buy the rest of the company as well.[63] Like Alco, Worthington was a recent casualty of poor management and declining sales. Worthington's earnings per share had fallen steadily from $6.35 in 1957 to $2.39 in 1963. The managers of both companies hoped that a merger would smooth out the "boom and bust" cycles associated with their respective product lines. As an analyst from *Forbes* explained, "There's a real possibility that the two convalescing cripples can help each other."[64] Effective December 31, 1964, the former Alco Products, Inc. became the Alco Products Division of the Worthington Corporation.

Unfortunately for Alco-Worthington, most railroads had completed their new motive power purchases by the end of 1966. This situation, combined with the temporary suspension of a 7 percent investment tax credit in early 1967, brought a quick end to the locomotive boom. Although railroads had placed orders for 1,204 locomotives in 1966, they planned to order only

166 the following year. Faced with increased competition from GE, Alco's market share began to decline: from 11 percent in 1966, to 6 percent the following year, to 3 percent in 1968. GE captured 33 percent of the market in 1968.[65]

The November 1967 merger of the Worthington Corporation with the Studebaker Corporation finally settled Alco's fate as a locomotive producer. The newly formed Studebaker-Worthington Corporation, with assets of some $550 million, failed to perform as well as was expected. In January 1969, after purging the company's top management, the board of directors appointed Derald H. Ruttenberg, a Chicago lawyer, as president and chief executive officer. Ruttenberg pledged to eliminate all of the financially unsound components of the corporation. First to go was the Alco Products Division. Locomotive orders were virtually nonexistent, and GE was once again in the process of expanding its line of diesel locomotives. Accordingly, Alco ended 121 years of continuous locomotive production at Schenectady in 1969. In February 1970, Studebaker-Worthington sold its diesel engine business to the White Motor Corporation. The Montreal Locomotive Works remained a component of Studebaker-Worthington and continued to build diesel locomotives to Alco designs, primarily for the Canadian market.[66]

Success and Failure at ALCo

For more than forty years, ALCo produced a variety of diesel locomotives and sold them to railroads throughout the United States and abroad, far longer than either Baldwin or Lima. ALCo's Canadian subsidiary, the Montreal Locomotive Works, significantly outperformed GM Diesel, Ltd. ALCo made impressive gains in its transformation from a steam locomotive builder to a diesel locomotive producer. After World War II, the board of directors authorized a major plant rationalization and rewarded, albeit belatedly, the managerial talents of diesel advocate Perry Egbert. The creation of the Century Series locomotives constituted a sound business decision, and efforts at diversification, while ultimately fruitless, were at least well intentioned.

These efforts were insufficient, however. In manufacturing, ALCo did not do enough to establish a fully integrated facility for the standardized assembly-line production of diesel locomotives. Having dispersed its financial resources through an excessive dividend policy in the 1920s, the company lacked the financial ability to transform its Schenectady plant into a modern diesel locomotive production facility, or to launch a more effective R&D program. Instead, ALCo produced these locomotives in a variety of unsuitable surplus buildings scattered throughout the plant site. Even the much-heralded postwar reconversion program did not produce an efficient plant comparable to EMD's La Grange facility. Furthermore, ALCo, throughout its career as a locomotive builder, relied on General Electric for vital and expensive electric equipment. Because ALCo had little control over the construction of one of the three main components of a diesel locomotive, the company lost some of its profit potential and was vulnerable to the vagaries of GE and other independent suppliers. The combination of an outdated manufacturing facility and reliance on GE equipment reduced the level of control that ALCo's managers could exercise over the design and production process and ultimately resulted in locomotives that were increasingly below the standards established by EMD.

Furthermore, factors beyond ALCo's control contributed to the company's longevity in the diesel locomotive industry. In an industry characterized by duopoly, ALCo survived so long as it was the second-best producer, even if it was not competitive against EMD. Once GE supplanted ALCo's secondary position, the latter company soon succumbed. ALCo's customers in the railroad industry wished to preserve some competition in locomotive production; whether that role belonged to ALCo or GE was, from their perspective, immaterial, so long as both companies offered comparable products. By the mid-1960s, the quality and reliability of GE's diesels had surpassed their ALCo counterparts, and railroads responded by rapidly switching allegiances.

Although to a lesser degree than their counterparts at Baldwin and Lima, ALCo executives never fully understood that the transition from steam to diesel locomotive production also involved a shift from small-scale custom production to near-mass-production techniques. Whether they understood this added dimension or not, ALCo's difficulties during the late 1950s and into the 1960s stemmed from an agonizing decision: should the company remain true to its *production* heritage and emphasize the small-batch customized fabrication of items other than diesel locomotives even as the capital goods sector of the economy declined, or should the company continue its *product* heritage by abandoning custom manufacturing in favor of the near-mass-production of diesel locomotives? In the final analysis, the indecisiveness of ALCo's managers ensured that the company did neither.

Conclusion

As the American locomotive industry made the transition from steam to diesel locomotive production, companies rose and fell in response to the challenges created by that radical technological discontinuity. Corresponding changes in technologies, production methods, marketing services, and managerial cultures so thoroughly transformed the locomotive builders that the locomotive industry of 1970 bore scant resemblance to that of 1930. Company executives responded to new technologies while preserving the viability of their established product lines. Managers manipulated technological change to suit their own interests and those of their companies and their stockholders. Executives did not operate in a vacuum, however, for they interacted with individuals in the railroad industry, in government, and among the general population.

The locomotive industry, as a case study, raises a number of broad historical issues. The government, particularly the federal government, had the power to foster technological development, encourage economic growth, and distort competitive patterns, but that power often worked in subtle ways. Legislation designed to reduce pollution and increase safety, such as the 1923 Kaufman Act, did encourage technological innovation. Both ALCo and GE produced locomotive components to satisfy the demand created by the Kaufman Act and similar laws, but these efforts during the 1920s were not closely connected to ALCo's own diesel line of the 1930s and later or to GE's U-series locomotives of the 1960s. Furthermore, this legislation had no effect on Electro-Motive since that company produced only railcars during the 1920s.

The greatest single impact of governmental action arose from U.S. Navy interest in the Model 201 diesel engine. Military demand diverted Charles Kettering and the GM Research Labs away from their unsuccessful explorations into automobile and truck diesel engines. Coincidently, the Model 201 engine, sized to fit inside a submarine, was also a near-perfect fit for a railroad locomotive. Without continual prodding by Harold Hamilton, Kettering, and Ralph Budd, however, it is unlikely that the Navy engine would have found an application in railroad locomotives.

The War Production Board manipulated the locomotive industry but did little to distort long-term competitive patterns. Prewar and postwar market structures were virtually identical. Electro-Motive suffered more than any other producer, since its market share declined during the war. ALCo and Baldwin temporarily benefited but could not preserve their wartime market share gains in the postwar period. It would be a mistake to accept at face

CONCLUSION   147

value the postwar claims of ALCo executives that the WPB had given Electro-Motive an unfair advantage during the war. During World War II, ALCo executives, like their counterparts at Baldwin, raised no objections to WPB policy. Only later, when ALCo had let its own competitive position in the locomotive industry slip, did the company's executives blame the WPB rather than admit to more pervasive problems.

While the WPB had little effect on the locomotive industry, the war itself proved immensely valuable to Electro-Motive. Wartime military demand ensured Electro-Motive's profitability for the first time, causing GM to reorganize Electro-Motive from a subsidiary to a division. The newly created division expanded manufacturing capacity, standardized production methods (to accompany earlier design standardization), and placed increased emphasis on stability rather than experimentation. These changes served Electro-Motive well in the postwar years.

The federal government's antitrust prosecution of GM and Electro-Motive during the 1960s had little effect, other than to point out serious deficiencies in antitrust policy. While economists and policy-makers today exhibit less interest in the concentration of economic power than in earlier decades, the story of the locomotive industry still has relevance to that issue. Some scholars, such as John Kenneth Galbraith and Alfred D. Chandler Jr., have assuaged fears of economic concentration by demonstrating that the entrance of large diversified firms into a new industry can increase competition, innovation, and overall economic growth. The evolution of the locomotive industry supports this explanation to an extent, since GM and GE did increase temporarily the number of participants in the locomotive industry. Indeed, for a brief period, five firms manufactured diesel locomotives—significantly more than the number of producers in the steam locomotive industry. This situation was temporary, however, since only two producers have survived.

Demand patterns, not legal action, ensured that monopoly did not emerge in the locomotive industry. Railroads purchased ALCo locomotives during the 1950s and 1960s and bought GE locomotives during the 1960s and 1970s, primarily to keep a second producer in the industry. Their subsidization bore fruit, resulting in continued technological innovations at EMD and, by the 1980s, at GE as well. In the case of the locomotive industry, government antitrust policy was not needed to preserve competition. However, it should be pointed out that efforts to keep a secondary producer in the market must have imposed a tremendous cost on the railroad industry, distorted competitive patterns, and uneconomically prolonged ALCo's survival as a locomotive producer.

The evolution of the locomotive industry also shows that established, successful firms experience extraordinary difficulties in adapting to technological change. ALCo, Baldwin, and Lima weathered the periodic peaks and troughs of the "boom-and-bust" steam locomotive industry. They produced quality products, often using the most advanced technology available, and they acquired worldwide reputations as suppliers of high-quality steam locomotives. The steam locomotive builders proved extraordinarily adept at managing incremental technological change, continually refining and improving their steam locomotive designs, often by working in close cooperation with railroad motive power officials who were fiercely loyal to "their" steam locomotive supplier.

With diesels, however, everything went wrong. Since diesel locomotives constituted an entirely new technology during the 1920s and 1930s, any company that hoped to compete successfully in that industry would have to make extensive plant modifications and invest substantial sums in R and D programs. The only way to amortize high R and D expenditures was through long production runs—not twenty locomotives for one railroad but a thousand locomotives for dozens of railroads. This in turn mandated that ALCo, Baldwin, and Lima break with a century of tradition and force railroad motive power officials out of the design process. In so doing, they risked alienating their allies in the railroad industry, but a larger problem overshadowed this one. Loyalty, as it turned out, was a very tenuous thing, restricted as it was to traditionalists in railroad motive power departments. Railroad directors, presidents, and financial officials cared more about cost savings and operational efficiencies than about the power and romance of the steam locomotive. These officials were willing to accept standard-design diesel locomotives, provided that their builder could offer a quality product coupled with performance guarantees, warranties, training programs, financing options, and rapid spare parts service. Electro-Motive, not ALCo, Baldwin, or Lima, proved best able to offer these things.

Corporate culture lay at the heart of the adaptability problem. Based on decades of training and experience, executives at ALCo, Baldwin, and Lima had considerable familiarity with steam locomotive technology. Industry executives logically saw less immediate risk in maintaining their existing product line and operational routines than in testing uncharted waters. The failure of early diesels to meet traditionally valued steam locomotive performance criteria served to reinforce this managerial aversion to diesels. Samuel Vauclain's autocratic control over Baldwin offered the most extreme example of this reliance on the steam locomotive and distrust of the diesel, but many of his contemporaries held similar attitudes. And, a corporate culture that brooked no opposition to steam was unlikely to reward advocates of diesel locomotive technology—as attested by the slow-moving career of ALCo executive Perry Egbert.

The story was very different at Electro-Motive, however. Harold Hamilton largely created Electro-Motive, and he understood the importance of marketing and post-sale support services. During his years in the automo-

CONCLUSION   149

bile industry, Hamilton developed a keen appreciation for the ability of the proponents of an old technology, based on the horse, to resist a new technology, motorized trucks. He understood that a wide variety of marketing initiatives, including training programs, financing, and other support services, were as important, if not more important, than the technology itself. He marshaled this understanding as he guided Electro-Motive to market dominance in the railcar industry during the 1920s. Significantly, Electro-Motive had no real mastery of railcar manufacturing and did not actually produce a product; it subcontracted engine production to Winton and carbody construction to Pullman, the St. Louis Car Company, and a variety of other firms. Electro-Motive succeeded in the railcar industry because the company backed its products with performance guarantees and warranties, offered rapid spare-parts service, and trained and supervised railroad operating and maintenance personnel.

When Electro-Motive joined the GM corporate family in 1930, Hamilton and his colleagues had considerable marketing expertise yet lacked the technological knowledge and the financial resources necessary to transfer their skills to the diesel locomotive industry. GM provided the necessary funds and technical know-how (ALCo and Baldwin could have done the same, particularly if they had been more restrained in their dividend policies during the 1920s). The marriage of Electro-Motive's marketing skills and GM's technological strengths led to success in the diesel locomotive industry during the 1930s; incremental improvements to established technology and marketing programs led to market dominance by 1940.

During the 1930s, Electro-Motive's corporate culture placed a premium on innovation and experimentation, and this allowed Hamilton to find a ready ally in GM's research director, Charles Kettering. Most important, Hamilton, his associates at Electro-Motive, certain technical experts at GM (including Kettering), and a few forward-looking railroad executives (such as the Burlington's Ralph Budd) were all true believers. They had as much confidence in dieselization as executives at ALCo, Baldwin, and Lima had in the long-term survival of the steam locomotive, if not more. This enthusiasm, set within the larger realm of Electro-Motive's corporate culture, enabled Hamilton and his allies to manipulate subtly the GM corporate hierarchy and pursue policies based on personal devotion rather than corporate strategy.

Despite the overall rigidity of organizational routines and corporate cultures, Electro-Motive adapted to changing market conditions. As World War II began, military demand for diesel engines (unanticipated by Hamilton in the 1930s) far exceeded railroad demand for locomotives and finally ensured Electro-Motive's profitability. GM officials understood that an experimental corporate culture was not amenable to stability and predictable profitability. As a result, GM not only expanded Electro-Motive's manufacturing facilities; it also assigned enough GM organization men to Electro-Motive to change fundamentally Electro-Motive's corporate culture. Electro-Motive did not, and quite possibly could not, have initiated this necessary change in its corporate culture. Instead, the impetus for change came from a larger entity that was not affected by Electro-Motive's organizational routines. This oversight capability may be one of the most important attributes of the large diversified corporation.

Although two large, diversified firms (GM and GE) came to control the diesel locomotive industry, this does not indicate that mass production firms are inherently superior to small-batch custom producers. Custom production worked well in the steam locomotive industry, and scholars such as Philip Scranton, Michael Piore, and Charles Sabel are entirely correct in their assertion that this type of manufacturing could work well for certain companies in certain industries. Still, production strategies are largely dependent on the characteristics of a given technology. Diesel locomotive technology proved far more cost effective when linked with standardized designs and standardized manufacturing techniques and, accordingly, near-mass production came to characterize the diesel locomotive industry. ALCo, Baldwin, and Lima, skilled at customized production, found it necessary to change both product lines and production strategies, and this dual transition proved immensely difficult and ultimately fatal.

The notion of technological determinism has fallen out of favor among historians, and indeed, there is little in the locomotive industry story to suggest that deterministic forces were at work. ALCo possessed the requisite elements of diesel locomotive technology, yet company executives chose not to make the organizational changes required to develop this technology. While a maladaptive corporate culture crippled the efforts of the locomotive producers to respond to massive and discontinuous technological change, this culture was in no sense deterministic. Human actors managed the steam locomotive builders, and those human actors were free to make choices. Corporate culture certainly affected their actions and may have limited their options, but it did not determine any particular outcome. Electro-Motive possessed more sophisticated diesel engine technology yet required individual actors, such as Harold Hamilton and Ralph Budd, to prod and manipulate that technology in the direction of diesel locomotives. Furthermore, Electro-Motive's success in the diesel locomotive industry resulted partly from random chance, particularly during the 1930s—and random behavior is the antithesis of technological determinism.

As social constructivists would expect, the interactions of various social groups affected the development of the locomotive industry. Likewise, the actor-network approach, which places greater emphasis on the complexity of the amalgamation of heterogeneous actors (including physical elements ranging from Cor-Ten steel to alternating-current traction motors) into a stable network, has considerable explanatory power, particularly since it bypasses much of the artificially simplified linear power relationships inherent in sociological theory. Still, the story of the locomotive industry provides little evidence to indicate the *primacy* of extensive social relationship networks over other factors. While many actors and social groups revolved around the axis that linked locomotive suppliers and railroad purchasers, few of these marginal groups had any significant impact on the development of diesel locomotive technology or on the evolution of the locomotive industry. Actors in government had little influence—the Kaufman Act had no effect on Electro-Motive, the War Production Board did not distort competitive patterns in the industry, and the Justice Department's efforts during the 1960s proved ultimately fruitless. The general public responded enthusiastically to the *Zephyr* and welcomed clean diesel passenger power, yet switchers and road freight diesels brought Electro-Motive both its profits and its market dominance. Social conditions, in the form of corporate cultures, held sway largely within the comparatively narrow confines of the locomotive producers themselves. Aside from a handful of influential railroad customers, outside actors and social groups exerted scant influence.

The organizational synthesis approach to firm and industry behavior explains a great deal about the history of the locomotive producers, yet it does not provide a complete frame of reference. Much of Electro-Motive's success resulted from its organizational strengths, particularly in its research and development and marketing efforts. Electro-Motive succeeded, in large measure, because it established first-mover advantage in production, marketing, and management, even though its competitors (most notably ALCo) built diesel locomotives years earlier.

Organizational synthesis does not adequately explain the ways in which companies developed (or failed to develop) these organizational strengths when they did. In particular, the organizational synthesis paradigm, much like technological determinism and neoclassical economic theory, gives too little weight to historical actors. ALCo and Baldwin suffered because their intelligent and well-trained managers erroneously decided that diesel locomotives were unlikely to offer the performance characteristics that they considered important. This conclusion was a highly individual one, based on personal values and deeply ingrained modes of behavior, yet it affected the operation of their respective organizations. At the same time, workers on the shop floor resisted the devaluation of their traditional skills (accounting, in part, for the strikes that plagued ALCo's postwar production), and this human dimension lay largely outside the parameters of organizational control. Furthermore, EMD quite probably would not have been a significant force in the diesel locomotive industry had not individuals such as Harold

Hamilton, Charles Kettering, and Ralph Budd advocated the adaptation of diesel submarine engines for railroad use.

Hamilton's ability to manipulate the GM corporate bureaucracy calls into question the omniscience of the large industrial corporation. Of course, the large multidivisional corporation intentionally allows its divisions and subsidiaries considerable latitude to explore promising market opportunities with little risk or obligation to the parent company, and even Alfred Sloan recognized the inherent tension between centralized managerial control and localized experimentation and inventiveness. While proponents of the organizational synthesis would doubtless accept this tension, they might be less comfortable with the notion that both customers and subsidiary managers could independently decide to explore opportunities that the head office did not consider promising and that these individuals could intentionally manipulate and even mislead top corporate officials in order to further their own private agendas. Hamilton's ultimate success in managing technological change arose from his access to "capital that wouldn't control"—in his ability to manipulate the GM corporate hierarchy to suit his own ends in order to achieve his vision of widespread dieselization.

The foregoing analysis suggests that no single factor—be it organizational expertise, deterministic technology, social patterns, a favorable legal and political environment, or the presence of an inventor, an entrepreneur, or a community of technological practitioners—led to the success of certain firms and the failure of others. ALCo had many of these elements in its favor, including a solid organizational structure, respected executives, social and legislative support (in the case of the niche market created by the Kaufman Act), and adequate technology; yet ALCo did not survive. Electro-Motive survived, and thrived, only because it had all of these elements in its favor, in the right measure, in the correct order, and then only because its timing was unintentionally perfect.

The historical example of the locomotive industry indicates that a well-developed, successful, and respected corporate culture can lead a company to great economic success—but also that that corporate culture can lead the same company to disaster. The recent financial difficulties of many large and traditionally prosperous American firms, including General Motors, should call into question the adaptability of managers, organizational routines, and core capabilities. The locomotive industry offers an example of the stunningly successful effort of an attacking firm and its small core of dedicated manager-engineers to exploit radical technological change and take full advantage of the opportunities offered by a large, diversified corporation. It also provides a disturbing picture of successful, highly trained managers who were so immersed in corporate operating routines and so successful at providing steadily improved products for their customers that they could not

CONCLUSION

see the threat posed by a radically new technology. The ultimate lesson of the locomotive industry is that technology often changes more rapidly than the people whose careers depend on it and that, to be exploited effectively, technological change must be accompanied by more than plant modernization, by more than research and development, by more than government action; it must be accompanied by fundamental changes in the hearts and minds of those who attempt to control it.

Notes

Introduction

1. For an analysis of the evocative overtones of steam locomotive technology, see Wyn Wachhorst, "An American Motif: The Steam Locomotive in the Collective Imagination," *Southwest Review* 72:4 (Autumn 1987): 440–54.

2. In spite of widespread popular interest in railroads in general and steam locomotives in particular, comparatively little historical research has been conducted on the American locomotive industry during the mid-twentieth century. The only recent scholarship concerning the diesel locomotive industry has come from Thomas G. Marx, in "Technological Change and the Theory of the Firm: The American Locomotive Industry, 1920–1955," *Business History Review* 50 (Spring 1976): 1–24, and "The Diesel-Electric Locomotive Industry: A Study in Market Failures" (unpublished Ph.D. diss., University of Pennsylvania, 1973). Marx, an economic historian, studies issues that are considerably different from those addressed here. He offers a narrowly focused analysis of government antitrust policy in relation to two suits filed by the Justice Department against General Motors during the 1960s. Marx is less concerned with corporate decision-making processes or overall managerial responses to technological change. His primary interest lies in the realm of prescriptive macroeconomics policy analysis. In addition, Marx did not have access to the vast wealth of company records relating to individual firms now available to historians. Most of the other secondary works that describe the locomotive industry are intended primarily for the railfan market. They contain many beautiful photographs and exhaustive amounts of detail concerning specific locomotive types, experimental models, and railroad assignments, but provide little historical analysis. To a large extent, these works suffer from a common failing in that their primary focus is on the product rather than on the process of production. Nevertheless, they sometimes provide information not readily available elsewhere. Three of the most useful of these books have been written by John F. Kirkland: *The Diesel Builders, Vol. 1: Fairbanks-Morse and Lima-Hamilton* (Glendale, California: Interurban Press, 1985); *The Diesel Builders, Vol. 2: American Locomotive Company and Montreal Locomotive Works* (Glendale, California: Interurban Press, 1989); and Kirkland, *Dawn of the Diesel Age* (Glendale, California: Interurban Press, 1983).

3. Alfred D. Chandler Jr., *Scale and Scope: The Dynamics of Industrial Capitalism* (Cambridge: Harvard University Press, 1990), 658–65.

4. Both GM and GE are still very much involved in diesel locomotive production and, as such, neither company has granted access to their corporate archives. Nevertheless, other sources provide a wealth of information on the activities of these two companies, particularly during the formative years of the locomotive industry.

5. For example, see Joseph Schumpeter, *Capitalism, Socialism, and Democracy* (New York: Harper, 1942) and Schumpeter, "The Creative Response in Economic History," *Journal of Economic History* 7 (November 1947): 147–59. For analyses of

technological cycles, see: William Abernathy and James Utterback, "Patterns of Industrial Innovation," *Technology Review*, 80:7 (1978): 40–47; Utterback and Fernando F. Suarez, "Innovation, Competition, and Industry Structure," *Research Policy* 22 (1993): 1–21; and Philip Anderson and Michael L. Tushman, "Managing Through Cycles of Technological Change," in *Managing Strategic Innovation and Change: A Collection of Readings*, eds. Michael L. Tushman and Philip Anderson (New York: Oxford University Press, 1997): 45–52.

6. Edward W. Constant II, "The Social Locus of Technological Practice: Community, System, or Organization?" in *The Social Construction of Technological Systems: New Directions in the Sociology and History of Technology*, eds. Wiebe E. Bijker, Thomas P. Hughes, and Trevor Pinch (Cambridge: MIT Press, 1987), 223–42. Alfred Chandler is perhaps the best-known advocate of the organization as the center of technological practice, while Thomas Hughes is one of the leading supporters of a system-based approach. For an excellent overview of the historical literature on recent trends in business history and the history of technology, see David A. Hounshell, "Hughesian History of Technology and Chandlerian Business History: Parallels, Departures, and Critics," *History and Technology* 12 (1995): 205–24.

7. Thomas Parke Hughes, "The Evolution of Large Technological Systems," in *The Social Construction of Technological Systems*, eds. Bijker, Hughes, and Pinch, 51–82, provides a succinct overview of the stages involved in the process of technological change. The economist Paul A. David also discusses the generation and diffusion of technology in a historical context in *Technical Choice, Innovation, and Economic Growth: Essays on American and British Experience in the Nineteenth Century* (London: Cambridge University Press, 1975).

8. Nathan Rosenberg, *Inside the Black Box: Technology and Economics* (Cambridge: Cambridge University Press, 1982) and *Exploring the Black Box: Technology, Economics, and History* (Cambridge: Cambridge University Press, 1994).

9. This view tends to support that of evolutionary economists such as Richard R. Nelson and Sidney G. Winter. For example, in *An Evolutionary Theory of Economic Change* (Cambridge: Harvard University Press, 1982), they "[recognize] that there are stochastic elements both in the determination of decisions and of decision outcomes" (p. 15). They also criticize orthodox economic theory for its inability to accommodate a "plain old mistake" (p. 8). As the following chapters will show, random events certainly affected the locomotive industry, and all of the locomotive producers had their share of plain old mistakes.

10. Edward W. Constant II, "The Social Locus of Technological Practice: Community, System, or Organization?" in *The Social Construction of Technological Systems*, eds. Bijker, Hughes, and Pinch, 223–42. Thomas P. Hughes, in *Networks of Power: Electrification in Western Society* (Baltimore: Johns Hopkins University Press, 1983), uses the more inclusive term "system builders."

11. There has been a long historiographical dispute over whether the lone inventor or the corporate R and D laboratory is the instigator of technological change. Recent work on Thomas Edison (the classic inventor-hero), including Robert Friedel and Paul Israel, *Edison's Electric Light: Biography of an Invention* (New Brunswick: Rutgers University Press, 1986), delineates Edison's role as a manager within a complex technical system. George Wise, "A New Role for Professional Scientists in Industry: Industrial Research at General Electric, 1900–1916," *Technology and Cul-*

ture 21 (July 1980): 408–29; David A. Hounshell and John Kenly Smith Jr., *Science and Corporate Strategy: DuPont R&D, 1902–1980* (Cambridge: Cambridge University Press, 1988); and Leonard S. Reich, *The Making of American Industrial Research: Science and Business at GE and Bell, 1876–1926* (Cambridge: Cambridge University Press, 1985), explore the growth of corporate R and D capabilities. Thomas Parke Hughes, *Elmer Sperry: Inventor and Engineer* (Baltimore: Johns Hopkins University Press, 1971), explores the tensions between a heroic inventor and the growing corporate control over invention and R and D. Steven W. Usselman, "From Novelty to Utility: George Westinghouse and the Business of Invention during the Age of Edison," *Business History Review* 66 (Summer 1992): 251–304, describes two distinctly different, although uniquely individualistic approaches, to the invention and diffusion of technology.

12. Christopher Freeman, *The Economics of Industrial Innovation*, 2d. ed. (London: Francis Pinter, 1982), provides an excellent theoretical discussion of patterns of innovation and innovation strategies. He distinguishes (p. 19) between product innovations, process innovations (of considerable importance to the locomotive industry), energy innovations, and materials innovations (also important, especially to Electro-Motive). See also Richard R. Nelson and Sidney G. Winter, "In Search of a Useful Theory of Innovation," *Research Policy* 6 (1977): 36–76.

13. Chandler, for example, has stressed that the mere possession of applicable technology, patented or not, does not guarantee success in any industry. As he makes clear in *Scale and Scope*, first-mover advantages accrue only to those companies that exploit their technology by enacting complementary process innovations in manufacturing, marketing, and management.

14. For good analyses of various forms of technological determinism and competing contextualist/social constructivist viewpoints, see: Steven W. Usselman, "Determining a Middle Landscape: Competing Narratives in the History of Technology," *Reviews in American History* 23 (1995): 370–77; and David B. Sicila, "Technological Determinism and the Firm," *Business and Economic History* 22:1 (Fall 1993): 67–78. Paul David studies path dependency in relationship to the typewriter in: "Clio and the Economics of QWERTY," *American Economic Review* 75:2 (May 1985): 332–37 and "Understanding the Economics of QWERTY: The Necessity of History," in *Economic History and the Modern Economist*, ed. William N. Parker (Oxford: Basil Blackwell, 1986), 30–49. The same author's "Heroes, Herds and Hysteresis in Technological History," *Industrial and Corporate Change* 1 (1992): 129–80, shows how "the personal ambitions and subjective beliefs of inner-directed entrepreneurs, and the particular ideological convictions of public policy-makers, become essential ingredients of the temporary historical contexts within which the long-run course of change may be pushed irreversibly in one direction rather than another." (P. 134.)

15. Chandler, in *Scale and Scope*, argues convincingly that patents are not a key determinant of firm success—a conclusion supported by this analysis of the locomotive industry. For additional studies on the role of patents and the diffusion of patented technology, see Steven W. Usselman, "Air Brakes for Freight Trains: Technological Innovation in the American Railroad Industry, 1869–1900," *Business History Review* 58 (Spring 1984): 30–50.

16. For an analysis of the operation of the patent system in relation to the railroad industry, see Steven W. Usselman, "Patents Purloined: Railroads, Inventors, and the

Diffusion of Innovation in 19th-Century America," *Technology and Culture* 32:4 (October 1991): 1047–75.

17. For a discussion of the interaction between business development and national economic performance, see William Lazonick, *Business Organization and the Myth of the Market Economy* (Cambridge: Cambridge University Press, 1991).

18. Giovanni Dosi, *Technical Change and Industrial Transformation: The Theory and an Application to the Semiconductor Industry* (London: Macmillan, 1984), 4–5, 98.

19. Schumpeter was one of the first economists to address the issue of economic concentration. Dosi, in *Technical Change and Industrial Transformation*, also explores this issue. Alfred D. Chandler Jr., *The Visible Hand: The Managerial Revolution in American Business* (Cambridge: Harvard University Press, 1977), examines the increasing control of corporate hierarchies over all aspects of business activity, but does not provide much detail on technology per se.

20. In his study of the American telephone industry, Louis Galambos rejects both mainstream neo-classical economic theory and the concept of technological determinism as explanations for industry structure. Instead, he argues that political considerations and governmental controls had a greater impact than markets or technology. (Galambos, "Looking for the Boundaries of Technological Determinism: A Brief History of the U.S. Telephone System," in *The Development of Large Technological Systems*, eds. Renate Mayntz and Thomas P. Hughes [Boulder, Colorado: Westview Press, 1988].) This was not the case in the locomotive industry—a conclusion considerably different from that developed by Thomas Marx, whose dissertation, "The Diesel-Electric Locomotive Industry: A Study in Market Failures" (the title effectively summarizes his conclusions), stresses the need for corrective government intervention, in part to remedy exogenous factors induced by past government intervention.

21. Chandler, *Strategy and Structure: Chapters in the History of American Industrial Enterprise* (Cambridge: MIT Press, 1962). Also, Chandler, *The Visible Hand* (1977) and *Scale and Scope* (1990). For scholarship regarding small-batch manufacturing, see: Philip Scranton, "Diversity in Diversity: Flexible Production and American Industrialization, 1880–1930," *Business History Review* 65 (Spring 1991): 27–90; Scranton, *Figured Tapestry: Production, Markets, and Power in Philadelphia Textiles, 1885–1941* (Cambridge: Cambridge University Press, 1989); and Michael J. Piore and Charles F. Sabel, *The Second Industrial Divide: Possibilities for Prosperity* (New York: Basic Books, 1984).

22. Michael L. Tushman and Philip Anderson, in "Technological Discontinuities and Organizational Environments," *Administrative Science Quarterly* 31 (September 1986): 439–65, discuss incremental (competence-enhancing) and radical (competence-destroying) technological changes in the context of the minicomputer, cement, and airline industries. They correctly label the diesel locomotive as a radical discontinuity. Rebecca M. Henderson and Kim B. Clark offer a more sophisticated analysis of technological discontinuities in "Architectural Innovation: The Reconfiguration of Existing Product Technologies and the Failure of Established Firms," *Administrative Science Quarterly* 35 (March 1990): 9–30. They add two further types of discontinuities; namely, modular innovation, which produces substantial changes in product components, but has little effect on the way in which these components fit to-

NOTES TO INTRODUCTION 159

gether into the product's "architecture," and architectural innovation, in which basic components remain essentially unchanged, but are put together in a new way. Henderson and Clark emphasize that architectural innovations, while seemingly minor, can require substantial changes in organizational routines, and thus can be nearly as devastating to a firm as radical discontinuities.

23. Kenneth Lipartito, in "Innovation, the Firm, and Society," *Business and Economic History* 22:1 (Fall 1993): 92–104, writes that "in light of the substantial inertia surrounding the innovation process, radical changes may require special types of actors . . . to break up existing patterns of thought or behavior. . . . Oligopolistic competition among successful firms in normally functioning markets where innovation is proceeding along incremental lines probably cannot induce such radical change" (p. 100). This condition applied to the steam locomotive industry.

24. Richard N. Foster, *Innovation: The Attacker's Advantage* (New York: Summit Books, 1986).

25. Clayton M. Christensen, "The Rigid Disk Drive Industry: A History of Commercial and Technological Turbulence, *Business History Review* 67 (Winter 1993): 531–88; Richard S. Rosenbloom and Christensen, "Technological Discontinuities, Organizational Capabilities, and Strategic Commitments," *Industrial and Corporate Change* 3 (1994): 655–85; and Christensen and Rosenbloom, "Explaining the Attacker's Advantage: Technological Paradigms, Organizational Dynamics, and the Value Network," *Research Policy* 24 (1995): 233–57.

26. The definition of corporate culture used here refers to the beliefs, attitudes, and values of company management, the way in which these beliefs were shaped by education and work experience, and the impact of the resulting values on corporate decision-making processes. It also encompasses the ways in which corporate executives were rewarded (or punished) for their conformity (or lack thereof) to these norms and standards.

27. Dorothy Leonard-Barton postulates that four key characteristics influence organizational core capabilities: employee knowledge and skills, technical systems, managerial systems, and values and norms. Leonard-Barton emphasizes that "all four dimensions of core capabilities reflect accumulated behaviors and beliefs based on early corporate success." The steam locomotive producers developed considerable skills in all four of these dimensions, yet all were inappropriate for the diesel locomotive industry. The last of these, in particular, proved quite difficult to modify. In the case of the locomotive industry, "values and norms," in the form of management's corporate culture, constrained the ability of these companies to adapt to radical technological change. Dorothy Leonard-Barton, "Core Capabilities and Core Rigidities: A Paradox in Managing New Product Development," in *Managing Strategic Innovation and Change: A Collection of Readings*, eds. Michael L. Tushman and Philip Anderson (New York: Oxford University Press, 1997): 255–70.

28. In his 1995 Newcomen Prize Essay, "Culture and the Practice of Business History," *Business and Economic History* 24:2 (Winter 1995): 1–41, Kenneth Lipartito points out the need for a cultural approach to business history: "Reducing business behavior to the pursuit of profit, growth, and stability cannot account for the divergence of competitors located in the same market, sharing the same technological constraints, facing the same government mandates. It seems at least possible that culture supplies one of the missing pieces to this puzzle" (p. 2). An early, although

still useful, broad analysis of the interrelationship between culture and society is Clifford Geertz, *The Interpretation of Cultures* (New York: Basic Books, 1973). Hundreds, if not thousands, of books and articles relating to corporate culture have appeared in the business press over the past decade. Perhaps the best known is Thomas Peters and Robert H. Waterman, *In Search of Excellence* (New York: Harper & Row, 1982). Several historians have explored links between culture, technological change, and business performance. For example, see: Charles Dellheim, "Business in Time: The Historian and Corporate Culture," *Public Historian* 8 (Spring 1986): 9–22; Dellheim, "The Creation of a Company Culture: *Cadburys*, 1861–1931," *American Historical Review* 92 (February 1987): 13–44; Michael Rowlinson and John Hassard, "The Invention of a Corporate Culture: A History of the Histories of Cadbury," *Human Relations* 46 (March 1993): 299–326; William R. Childs, "The Transformation of the Railroad Commission of Texas, 1917–1940: Business-Government Relations and the Importance of Personality, Agency Culture, and Regional Differences," *Business History Review* 65 (Summer 1991): 285–344; and David M. Vrooman, *Daniel Willard and Progressive Management on the Baltimore & Ohio Railroad* (Columbus: Ohio State University Press, 1991). David A. Hounshell, "Elisha Gray and the Telephone: On the Disadvantages of Being an Expert," *Technology and Culture* 16 (April 1975): 133–61, shows how the "other" inventor of the telephone was so much a part of "a community of experts" (p. 159) that he was blinded to a wholly different technological paradigm, one that recognized the value of the telephone as more than a mere curiosity. Louis Galambos, "Theodore N. Vail and the Role of Innovation in the Modern Bell System," *Business History Review* 66 (Spring 1992): 95–126, provides a more positive view of one individual's ability to shape a viable corporate (or system) culture as a method of managing technological change. On a broader level, Thomas Hughes, in *Networks of Power*, discusses "system culture," which he defines as the "values, ideas, and institutions that arose in response to the needs and opportunities for growth that were seen and defined by professional engineers, managers, and system builders. . . ." (p. 363).

29. Giovanni Dosi uses the phrase "technological paradigm" to describe "a 'pattern' of solution of *selected* technological problems." Dosi uses the term "technological trajectory" to refer to technological advancements made within a given technological paradigm. (Dosi, *Technological Change and Industrial Transformation*, 14–15.) Nelson and Winter use the similar "technological regime." Constant, in *The Origins of the Turbojet Revolution*, uses the phrase "technological tradition." The Hughesian notion of "technological style" is also relevant here, although it relates to larger entities (such as nations) than those explored in this study. The social constructivist scholar Wiebe E. Bijker uses "technological frame," which he defines as "the concepts and techniques employed by a community in its problem solving." (Bijker, "The Social Construction of Bakelite: Toward a Theory of Invention," in *The Social Construction of Technological Systems*, eds. Bijker, Hughes, and Pinch, 159–87; and Trevor Pinch and Bijker, "The Social Construction of Facts and Artifacts: Or How the Sociology of Science and the Sociology of Technology Might Benefit Each Other," same volume, 17–50.) The social constructivist approach, like the Hughesian system-builder framework, emphasizes the complex interaction of social groups in the shaping of technology. The related actor-network paradigm asserts that "technological form rests on . . . both the *conditions* and the *tactics* of system building." In

other words, technology is shaped by the amalgamation of dissimilar physical and social elements into a self-sustaining network (John Law, "Technology and Heterogeneous Engineering: The Case of Portuguese Expansion," in Bijker, 111–34). See also Michel Callon, "Society in the Making: The Study of Technology as a Tool for Sociological Analysis," same volume, 83–103, and Bruno Latour, *Science in Action: How to Follow Scientists and Engineers through Society* (Cambridge: Harvard University Press, 1987).

30. This is Edward Constant's "presumptive anomaly"—the ability to predict that functioning technological systems will eventually reach the outer limits of their capabilities and will thus be replaced by radically different technology. Constant, *The Origins of the Turbojet Revolution* (Baltimore: Johns Hopkins University Press, 1980). Freeman discusses the related issue of "technical forecasting" and the various barriers to successful forecasting and to the integration of those forecasts into normal operating routines. Freeman, *The Economics of Industrial Innovation*, 2d ed., 165–67.

31. For an analysis of the relationship between the diesel locomotive (as replacement technology) and social change, with particular reference to railroad labor, see Maury Klein, "Replacement Technology: The Diesel as a Case Study," *Railroad History* 162 (Spring 1990): 109–20.

32. Nelson and Winter use clusters of organizational routines to explain firm behavior in *An Evolutionary Theory of Economic Change*. See also: Nelson and Winter, "In Search of a Useful Theory of Innovation."

Chapter I
Steam vs. Diesel

1. *Barron's*, June 7, 1937, 9.

2. The American railroad industry maintained close connections with all of its equipment suppliers, not just the locomotive builders. Steven W. Usselman gives another example of this collaboration in "Air Brakes for Freight Trains."

3. John K. Brown, *The Baldwin Locomotive Works, 1831–1915* (Baltimore: Johns Hopkins University Press, 1995), 230. The British locomotive industry also exhibited a multiplicity of steam locomotive designs, despite the fact that most large railroads built their own locomotives. See M. W. Kirby, "Product Proliferation in the British Locomotive Building Industry, 1850–1914: An Engineer's Paradise?" *Business History* 30:3 (July 1988): 287–305.

4. For a fascinating contemporary (1880s) account of steam locomotive production practices, see John K. Brown and Samuel Vauclain, "Comments on the System and Shop Practices of the Baldwin Locomotive Works, *Railroad History* 173 (Autumn 1995): 102–41. David Weitzman, *Superpower: The Making of a Steam Locomotive* (Boston: David R. Godine, 1987), offers a highly readable and visually impressive account of steam locomotive construction at Lima, although all three builders employed virtually identical techniques.

5. Earle Davis, "60 Years of Lima Locomotives," *Railroad Magazine*, December 1939, 25.

6. *Clayton Act*, Statutes at Large, 38, 323 (1914).

7. Charles A. Acker to Charles E. Denney, December 6, 1940, Northern Pacific

Railway Records, President's Subject Files, Minnesota Historical Society, St. Paul (hereafter referred to as the NP Collection), box 828, file 2241–40, part 1, 137.F.15.8(F).

8. Norman C. Naylor to Denney, December 5, 1940, NP Collection, box 828, file 2241–40, part 1, 137.F.15.8(F).

9. Brown, *The Baldwin Locomotive Works*, 170–83.

10. For a comprehensive discussion of customer-driven innovation in the steam locomotive industry, see Brown, *The Baldwin Locomotive Works*, 57–91.

11. *Railway Age* 88:1 (January 4, 1930), 79.

12. The diesel locomotive thus offers a clear example of a radical technological discontinuity—"no increase in scale, efficiency, or design can make older technologies competitive with the new technology." (Tushman and Anderson, "Technological Discontinuities and Organizational Environments," 441.) For a modern comparison of steam and diesel locomotive operating capabilities, see Robert Aldag, "Steam vs. Diesel Locomotives," *Railroad History* 167 (Autumn 1992): 148–57.

13. For the first half century after its invention, Dr. Rudolf Diesel's engine was usually referred to as the Diesel (capitalized) engine in his honor. More recently, it has been commonly referred to as the diesel (uncapitalized) engine. Except in the case of direct quotations, this work will use the latter version throughout. Even though railroad locomotives are often called "engines," the phrase "diesel engine" refers to the power plant alone, while a "diesel locomotive" includes the electrical equipment, carbody, underframe, trucks, and other components necessary for railroad use.

14. For more information on the career of Rudolf Diesel, see Donald E. Thomas Jr., *Diesel: Technology and Society in Industrial Germany* (Tuscaloosa: University of Alabama Press, 1987). This book, part biography and part history of technology, discusses the growth of a new profession, engineering, and the resistance to this by older established professions. It also examines the role of engineers as agents of change and solvers of social problems. Three articles by Lynwood Bryant are also useful: "The Development of the Diesel Engine," *Technology and Culture* 17 (July 1976): 432–46; "The Internal Combustion Engine," in *A History of Technology, Vol. VII: The Twentieth Century, c. 1900 to c. 1950, Part II*, ed. Trevor I. Williams (Oxford: Clarendon Press, 1978), 997–1024; and "Rudolf Diesel and his Rational Engine," *Scientific American* 221 (August 1969): 108–17; and Eugen Diesel, "Rudolf Diesel and his Rational Engine," *Engineering* 167 (March 18, 1949), 257, 281–82.

15. Diesel's 1892 patent was predated by an 1890 patent, issued to Herbert Ackroyd Stewart, for a semi-diesel engine. Unlike a true diesel, this engine required the fuel to be heated before it was injected into the cylinder. Several British scholars, with perhaps a touch of patriotism, have suggested that this British engine represented the true origins of the diesel. C. Lyle Cummins Jr., *Diesel's Engine: Volume One: From Conception to 1918* (Wilsonville, Oregon: Carnot Press, 1993), 466–67; Donal [sic] Borland, "Research Report TI-8: Diesel Development at the GM Research Laboratories, 1920–1938" (Warren, Michigan: GM Research Labs, 1967), General Motors Institute Alumni Foundation's Collection of Industrial History, Flint, Michigan (hereafter referred to as GMI); *Scientific American* 221:2 (August 1969), 108–17; Nelson C. Dezendorf, "Diesel Engines or Gas Turbines for Locomotives?" paper presented to the Pan-American Railway Congress, Mexico City, Octo-

ber 1950, Association of American Railroads Library, Washington (hereafter referred to as AAR).

16. Kirkland, *Dawn of the Diesel Age*, 32, 36; Nelson C. Dezendorf, address at the annual dinner of the Traffic Club of St. Louis, St. Louis, January 15, 1947, AAR.

17. Dezendorf, "Diesel Engines or Gas Turbines for Locomotives?" *Fortune* 38 (July 1948), 76–81, 144–49.

18. Thomas Hughes, in *Networks of Power*, has applied the concept of reverse salients, originally military jargon, to the study of the history of technology. Rather like an inward bulge in a line of battle, the reverse salient is a bottleneck, or a weak link in a chain—the least efficient technology, device, or production process that constrains the performance of an organization. In the context of this study, steam locomotives constituted the most obvious and pervasive source of inefficiency throughout the American railroad network by the end of the 1920s.

19. Garmany, *Southern Pacific Dieselization*, 55.

20. *Railway Age* 79:5 (August 1, 1925), 209.

21. *Railway Age* 74:3 (January 20, 1923), 241–43; 78:19 (April 11, 1925), 939–41; 88:19 (May 10, 1930), 1135; *General Electric Review* 40:9 (September 1937), 421–28 (quote).

22. For example, the diesel locomotives placed in service in 1948 alone reduced railroad coal consumption by ten million tons from the previous year. By 1954, diesels were saving railroads more than $600 million per year in fuel costs.

23. *Railway Age* 121:4 (July 27, 1946), 125; National Coal Association, "Memorandum for Members: Subject: The Growth of Diesel Locomotives," January 24, 1949, AAR; Harold L. Hamilton, speech at luncheon given by Alfred P. Sloan Jr., New York, October 28, 1937, AAR; Nelson C. Dezendorf, speech before the New York Society of Security Analysts, New York, June 1, 1951, AAR; Dezendorf, "Remarks at the formal opening of the enlarged Jacksonville, Florida, rebuild plant of Electro-Motive Division," Jacksonville, March 18, 1954, AAR.

24. *Railway Age*, 118:12 (March 24, 1945), 541–42; Remarks of Ralph Budd at a luncheon given by Alfred P. Sloan Jr., New York, October 28, 1937, AAR.

25. *Railway Age* 123:12 (September 20, 1947), 476–77; *Railway Progress* 12:2 (April 1958), 32–43; Hamilton, remarks at luncheon given by Alfred P. Sloan Jr., New York, October 28, 1937, AAR; GM-EMD, *Fuel Oil for Diesel Locomotives*, September 25, 1948, GMI, folder 19/26; *General Electric Review* 40:9 (September 1937), 421–28; American Locomotive Company, "Report of the Annual Meeting," April 19, 1949, The American Locomotive Company Collection at the George Arents Research Library, Syracuse University (hereafter referred to as the ALCo Collection), box 6; GM Research Laboratories Division, *Fuel Oil for Diesel Locomotives in Relation to the Supply of Petroleum*, September 25, 1948, AAR.

26. *Railway Age* 123:20 (November 15, 1947), 829–31; GM-EMD, "Why America Needs *More* Diesels *Now*," 1950, AAR; GM-EMD, "Conference Leader's Outline, Management Conferences, 1944–1945," subject II, units 1 & 3, GMI, folder 76–1.59; Dezendorf, speech before the New York Society of Security Analysts, New York, June 1, 1951, AAR; Diesel Engine Manufacturers Association, *Diesel Locomotives*, 1945, AAR.

27. Because tractive effort represents the actual force available to move trains, it offers a more useful indication of locomotive performance than does horsepower.

Tractive effort equals the total weight of the locomotive on its driving wheels multiplied by a coefficient of adhesion. The latter figure depends on the condition of the rail (i.e., wet, dry, oily, etc.) and varies little between steam and diesel locomotives. However, since steam locomotives carry much of their weight on unpowered lead and trailing trucks (necessary for the negotiation of curves and switches), the former figure is lower than that of diesels, which generally carry all of their weight on powered wheels (although some builders offered models with unpowered axles, in order to achieve better weight distribution). John A. Armstrong, *The Railroad: What It Is, What It Does* (Omaha: The Simmons-Boardman Publishing Corporation, 1982).

28. *Railway Age* 96:19 (May 12, 1934), 685–89; *Railway Progress* 12:2 (April 1958), 32–43; "Super Power for Super-Railroads," address by John W. Barriger III, before the Committee of the Coordinated Railway Mechanical Associations, Chicago, September 15, 1947, AAR.

29. Diesel Engine Manufacturers Association, *Diesel Locomotives*, 1945, AAR; GM-EMD, "Why America Needs *More* Diesels *Now*," 1950, AAR; EMD, "Conference Leader's Outline," subject III, unit 2B, GMI, folder 76–1.60.

30. For example, at high speeds, the drivers of a large steam locomotive could bounce as much as 1–1/4 inches off the rail. Franklin M. Reck, *On Time: The History of Electro-Motive Division of General Motors Corporation* (GM-EMD, 1948), 104.

31. *Railway Age* 100:19 (May 9, 1936), 763–66; 121:13 (September 28, 1946), 513; E. E. Chapman, "Operation of Steam vs. Diesel-Electric Locomotives," presented at the meeting of the American Society of Mechanical Engineers, Kansas City, Missouri, June 16–19, 1941, AAR; Diesel Engine Manufacturers Association, *Diesel Locomotives*, 1945, AAR; EMD, "Conference Leader's Outline," subject II, unit 3, GMI, folder 76–1.59; Dezendorf, speech before the New York Society of Security Analysts, New York, June 1, 1951, AAR.

32. *Railway Age* 102:10 (March 6, 1937), 402.

33. In recent years railroads have asserted that their entire industry is in danger of financial collapse and have won significant concessions from operating unions. These agreements now allow most long-distance trains to be operated by two crew members. The potential labor cost savings associated with dieselization have finally been realized, albeit some forty years after the disappearance of the last steam locomotives.

34. Electro-Motive average is based on models in production from 1936 to 1960. Diesel locomotive builders offered a variety of optional features, ranging from dynamic brakes to steam boilers for passenger train heating, but these altered the standardized designs only slightly.

35. W. R. Elsey memorandum, July 5, 1944, The Pennsylvania Railroad Collection at the Hagley Museum and Library, Wilmington, Delaware (MSS 1810, hereafter referred to as the PRR Collection), box 597, folder 416/11.

36. J. K. Thompson to Charles E. Denney, December 9, 1940, NP Collection, box 828, file 2241–40, part 1, 137.F.15.8(F).

37. Christensen, "The Rigid Disk Drive Industry," demonstrates the effects of new performance characteristics (a new "value network") on the data storage industry. Established manufacturers felt secure in their ability to manage technological change, since they increased storage capacity faster than their customers demanded.

NOTES TO CHAPTER II

They did not see smaller disk drives as a threat, since these initially offered greatly reduced storage capacity. Instead, new entrants succeeded because these smaller drives offered different performance capabilities (compactness, ruggedness, reduced vibration) that attracted new market segments. As demand for smaller disk drives increased, so too did their storage capacity, enabling new entrants to invade higher-end markets and compete effectively against established producers of large-drive systems.

Chapter II
Internal-Combustion Railcars

1. Railway equipment producers generally differentiated between "railcars" and "railbuses." The former were large, more powerful, and better suited to long branch line or mainline runs. Railbuses were much smaller and were often, as their name suggests, simply highway buses modified to fit on railroad wheels. While several companies produced railbuses commercially, financially strapped railroads often created their own homemade versions. Railbuses provided minimal service on lightly traveled branch lines, and usually lacked toilets and the other basic passenger amenities found on railcars.

2. Streetcars employed an external electrical power source, almost always an overhead wire, for propulsion and restricted their operations to urban areas. Interurbans drew power from either overhead wires or a ground-level third rail, were usually larger than streetcars, operated at higher speeds, and interconnected various urban and suburban areas. Straight electric locomotives did not themselves carry freight or passengers, but instead hauled separate railroad cars in mainline service. Although straight electric locomotives were more efficient than either steam or diesel locomotives, the enormous capital expenditure required to erect and maintain catenary (overhead power transmission systems) typically restricted straight electrics to a few areas of extremely high traffic density.

3. *Railway Progress* 12:2 (April 1958), 32–43; Stanley Berge and Donald L. Loftus, *Diesel Motor Trains: An Economic Evaluation* (Chicago: The Northwestern University School of Commerce, 1949), 2–4; Garmany, *Southern Pacific Dieselization*, 20, 25–30.

4. *Railway Age* 142:16 (April 22, 1957), 32–36, 42.

5. Cummins, *Diesel's Engine*, 695–98.

6. "Statement by Harold L. Hamilton," Congress, Senate, Committee on the Judiciary, Subcommittee on Antitrust and Monopoly, *A Study of the Antitrust Laws: Hearings before the Subcommittee on Antitrust and Monopoly of the Committee on the Judiciary*, 84th Cong., 1st sess., 1955 (hereafter referred to as the Senate Hearings), November 10, 1955, 2403; *Railway Age* 75:14 (October 6, 1923), 633–34; Berge and Loftus, *Diesel Motor Trains*, 2–3; Kirkland, *Dawn of the Diesel Age*, 67, 71–73. Garmany, *Southern Pacific Dieselization*, 33–34, 53.

7. Berge and Loftus, *Diesel Motor Trains*, 3; Reck, *On Time*, 16.

8. Cummins, *Diesel's Engine*, 695–98.

9. "Statement by Harold L. Hamilton," Senate Hearings, November 10, 1955, 2402–403. Hamilton recalled that "there were people in General Electric . . . who still had faith in the whole program. . . ."

10. *Railway Age* 132:15 (April 14, 1952), 57–58; 143:6 (August 5, 1957), 10; *Railway Progress* 12:2 (April 1958), 32–43; Franklin M. Reck, *The Dilworth Story* (New York: McGraw-Hill, 1954), 6, 10, 17–20, 26; Reck, *On Time*, 26.

11. For a detailed account of this process, see Carl W. Condit, *The Port of New York: A History of the Rail and Terminal System from the Grand Central Electrification to the Present* (Chicago: University of Chicago Press, 1981).

12. Martin Clement to John Deasy, November 29, 1927; Deasy to Fred W. Jankins; both in PRR Collection, box 334, file 416/15. Chicago passed similar legislation in 1912, but gave railroads until 1927 to comply, a time limit that the city later extended to 1935.

13. For additional studies concerning the effect of public concern for safety on the interaction between government and the railroad industry, see Mark Aldrich, "Combating the Collision Horror: The Interstate Commerce Commission and Automatic Train Control, 1900–1939," *Technology and Culture* 34 (January 1993): 49–77; and Steven W. Usselman, "The Lure of Technology and the Appeal of Order: Railroad Safety Regulation in Nineteenth Century America," *Business and Economic History* 21 (1992): 290–99.

14. In 1921, Ingersoll-Rand acquired the patent rights to the P-R engine, named for its developers, William T. Price, who later worked with Baldwin's subsidiary, De La Vergne, and George J. Rathburn. I-R used a modified version of this engine in early consortium locomotives. *Railway Age* 96:19 (May 12, 1934), 685–89; Kirkland, *Dawn of the Diesel Age*, 37–38, 76.

15. Condit, *The Port of New York*, 12.

16. ALCo's limited role as a locomotive subcontractor was illustrated by a New York Central order for forty-four locomotives, the largest single diesel locomotive order prior to 1946. GE and Ingersoll-Rand built these locomotives cooperatively, completely bypassing ALCo. *Railway Age* 76:23 (May 10, 1924), 1159; 83:19 (November 5, 1927), 890–91; 85:3 (July 21, 1928), 98–100; 86:12 (March 23, 1929), 663–67; Garmany, *Southern Pacific Dieselization*, 74.

17. Kirkland, *Dawn of the Diesel Age*, 75, 86–88, 97.

18. In 1937, Ingersoll-Rand withdrew from the railroad diesel engine market, having provided engines for 118 locomotives and two railcars.

19. *Railway Age* 86:26 (June 29, 1929), 1618; 97:7 (August 18, 1934), 209–10; Garmany, *Southern Pacific Dieselization*, 74; Kirkland, *Dawn of the Diesel Age*, 101, 153, 187.

20. *Railway Age* 111:18 (November 1, 1941), 724–25.

21. W. D. Bearce, "The Steam-electric Locomotive for the Union Pacific Railroad," *General Electric Review* 42:2 (February 1939): 87–91.

22. *General Electric Review* 40:9 (September 1937), 421–28 (quote); *Railway Age* 100:16 (April 18, 1936), 647–51; 101:18 (October 31, 1936), 615–21; 104:7 (February 12, 1938), 311; 109:21 (November 23, 1940), 784–86; 110:12 (March 22, 1941), 520–22; 119:3 (July 21, 1945), 115.

23. For a thorough analysis on how one railroad, the Southern Pacific, attempted to cope, albeit inadequately, with increased competition in passenger traffic during the interwar period, see Gregory Thompson, "Misused Product Costing in the American Railroad Industry: Southern Pacific Passenger Service between the Wars," *Business History Review* 63 (Autumn 1989): 510–54, and Thompson, *The Passenger Train*

in the Motor Age: California's Rail and Bus Industries, 1910–1941 (Columbus: Ohio State University Press, 1993).

24. Westinghouse Electric and Manufacturing Company, *Forty Years Ago: Being a Brief Account of the History and Accomplishments of the Westinghouse Electric and Manufacturing Company* (East Pittsburgh: Westinghouse, 1924); "Baldwin-Westinghouse Mine and Industrial Locomotives and Locomotive Equipment," Catalog #1774, May, 1927, AAR; Brown, *The Baldwin Locomotive Works*, 122.

25. For additional information regarding Westinghouse railroad equipment during the early years of the twentieth century, see: Westinghouse Electric & Manufacturing Company, *Cars and Car Equipment*, no date; George Westinghouse, "The Electrification of Railways: An Imperative Need for the Selection of a System for Universal Use," a paper prepared for the Joint Meeting of the American Society of Mechanical Engineers and the Institute of Mechanical Engineers, London, July 1910; and B. G. Lamme, *The Use of Alternating-Current for Heavy Railway Service* (Pittsburgh: Westinghouse Electric and Manufacturing Company, 1906); all in the Eugene Woodruff Industrial Electrification Papers, Historical Collections and Labor Archives, Pattee Library, The Pennsylvania State University.

26. *Baldwin Locomotives* 5:1 (July 1926), 43–49; 18:3+4 (February 1940), 31–33; *Railway Age* 79:15 (October 10, 1925), 645–48; Garmany, *Southern Pacific Dieselization*, 73; Kirkland, *Dawn of the Diesel Age*, 85–86.

27. *Railway Age* 79:17 (October 24, 1925), 757–58; 80:2 (January 9, 1926), 168–70; 80:17 (March 27, 1926), 911–13.

28. Ibid., 84:23 (June 9, 1928), 1319–23.

29. Westinghouse Electric and Manufacturing Company, "The Diesel Electric Rail Car . . . A Proved Transportation Tool," 1932, PRR Collection, box 598, file 9.

30. *Railway Age* 84:25 (June 23, 1928), 1451–54.

31. Ibid., 85:23 (December 8, 1928), 1125–27.

32. G. Maertz to W. W. Atterbury, December 11, 1928; J. H. Harvey, PRR Collection, box 598, file 8; *Railway Age* 86:14 (April 6, 1929), 787–90; 88:24 (June 14, 1930), 1427–29.

33. *Railway Age* 89:25 (December 20, 1930), 1347.

34. W. R. Stinemetz to F. W. Hankins, December 7, 1931, PRR Collection, box 598, file 9.

35. For example, one Beardmore-equipped unit was unavailable for service a total of twenty-eight days during a four-month period. F. W. Hankins to C. E. Smith, March 5, 1930, PRR Collection, box 477, file 415.3/6; W. F. Kiesel memorandum, November 28, 1934, PRR Collection, box 598, file 9; report on "Westinghouse oil-electric engine No. 23," ca. 1930, PRR Collection, box 598, file 8.

36. Harold Hamilton, president of Electro-Motive, felt that "the [Beardmore] engine was built by watchmakers apparently, and the minute something happened . . . you just about had to have that individual watchmaker on the job that built the first one the first time in order to get the engine back in shape." Expressing, perhaps, a touch of post–World War II patriotism, Hamilton criticized a similar German diesel engine design by stating that "it was so involved with so many auxiliary 'gadgets' and such refined workmanship and design that to manufacture it at any reasonable price—the way to make things in America—it was an impossibility." "Statement by Harold L. Hamilton," Senate Hearings, November 10, 1955, 2418.

37. "Features of Westinghouse 800 hp. Diesel Electric Locomotive," PRR Collection, box 598, file 9; *Railway Age* 80:7 (February 13, 1926), 443–46; 84:23 (June 9, 1928), 1319–23; 86:14 (April 6, 1929), 787–90; 88:24 (June 14, 1930), 1427–29; 95:5 (July 29, 1933), 179–81; 95:16 (October 14, 1933), 531–33; 96:19 (May 12, 1934), 685–89; 97:7 (August 18, 1934), 199–200; Garmany, *Southern Pacific Dieselization*, 73; Kirkland, *Dawn of the Diesel Age*, 109–11, 119, Jerry A. Pinkepank, *The Second Diesel Spotter's Guide* (Milwaukee: Kalmbach Publishing, 1973, 410–11.

38. Westinghouse, "The Diesel Electric Rail Car," PRR Collection, box 598, file 9.

39. *Railway Age* 97:7 (August 18, 1934), 199–200; 98:26 (June 29, 1935), 1044–45; 99:20 (November 16, 1935), 637–39.

40. Ibid., 98:1 (January 5, 1935), 17–19; 100:10 (March 7, 1936), 397–400.

41. Samuel M. Vauclain to W. W. Atterbury, January 21, 1929; John Van Buren Duer to Fred W. Hankins, June 5, 1935; both in PRR Collection, box 597, file 416/8.

42. By 1948, Westinghouse owned 22 percent of the outstanding Baldwin Stock. *Railway Age* 125:6 (August 7, 1948), 303; Baldwin press release, July 29, 1948, AAR; Kirkland, *Dawn of the Diesel Age*, 120, 126–27, 188.

43. The initials "EMC" refer to Electro-Motive from its origins in 1922 until its reorganization as the Electro-Motive Division of General Motors (EMD) on January 1, 1941. Throughout this book, "Electro-Motive" refers to the broad span of that organization's activities.

44. White sold approximately fifteen railbuses in all, many of which experienced problems with their mechanical transmissions. *Railway Age* 115:6 (August 7, 1943), 239–40; Reck, *On Time*, 11–14.

45. *Railway Age* 132:11 (March 17, 1952), 90–91; 132:15 (April 14, 1952), 57–58; *Railway Progress* 12:2 (April 1958), 32–43; Harold L. Hamilton, "Historical Background and Notes on the Development of Electro-Motive," November 22, 1946, GMI, folder 76–16.1; Reck, *On Time*, 14, 22; Reck, *The Dilworth Story*, 32.

46. Berge and Loftus, *Diesel Motor Trains: An Economic Evaluation*, 7; Reck, *On Time*, 59.

47. At GE, Dilworth had been a floor foreman in charge of experimental projects between 1912 and 1914, and he worked closely with Hermann Lemp on electrical power transmission equipment development. Cummins, *Diesel's Engine*, 706.

48. Reck, *The Dilworth Story*, 25–45; "Statement of Harold L. Hamilton," Senate Hearings, November 10, 1955, 2403 (quotes).

49. The Winton Engine Company and the better-known Winton Automobile Company were separate corporate entities, although both were founded by Alexander Winton. Hamilton, "Historical Background and Notes on the Development of Electro-Motive," 19–20, GMI; Reck, *On Time*, 58–60; Reck, *The Dilworth Story*, 39; Berge and Loftus, *Diesel Motor Trains*, 9–10.

50. "Statement of Harold L. Hamilton," Senate Hearings, November 10, 1955, 2405–6. Alexander Winton's pioneer spirit, along with the "faith" of GE employees in railcar technology, would seem to support Kenneth Lipartito's assertion that innovative firms "have no choice but to look ... to the enthusiasms of their best personnel, for they are projecting well beyond what can be known." Lipartito, "Culture and the Practice of Business History," 29.

51. Hamilton, "Historical Background and Notes on the Development of Electro-Motive," 19–20; Reck, *On Time*, 58–60; Reck, *The Dilworth Story*, 39; Berge and Loftus, *Diesel Motor Trains*, 9–10.

52. Clyde W. Truxell interview, The Kettering Archives, 1965 Oral History Project, March 10, 1961, 3–5; "History of the Winton Engine Company," ca. 1930, GMI, folder 76–16.1; Borland, "Research Report TI-8," 37, GMI; EMD *Streamliner* 12:2 (June 18, 1948), GMI, folder 83–12.101; *Diesel Times* 5:5 (May 1948), 1–8.

53. Hamilton, "Historical Background and Notes on the Development of Electro-Motive," 11.

54. National Railway Historical Society, Lehigh Valley Chapter, *History of Mack Rail Motor Cars and Locomotives* (Allentown, Pennsylvania: Lehigh Valley Chapter, National Railway Historical Society, 1959), 3; Berge and Loftus, *Diesel Motor Trains*, 5; Hamilton, "Historical Background and Notes on the Development of Electro-Motive," 14–15 (quote).

55. "Statement of Harold L. Hamilton," Senate Hearings, November 10, 1955, 2456.

56. Reck, *On Time*, 44.

57. EMC also provided financial incentives as part of its marketing efforts, since the company rented railcars to railroads unable to afford the purchase price. EMC specified that rental payments would be lower than the operating savings that the railcars were expected to create. Hamilton, "Historical Background and Notes on the Development of Electro-Motive," 22.

58. Ibid., 12; Reck, *On Time*, 50–52.

59. Hamilton, "Historical Background and Notes on the Development of Electro-Motive," 19; Reck, *On Time*, 56.

60. "Electro-Motive Power Plants Sold as of June 15, 1930," PRR Collection, box 519, folder 400.3/2.

61. Berge and Loftus, *Diesel Motor Trains*, 5; Hamilton, "Historical Background and Notes on the Development of Electro-Motive," 23.

62. Hamilton, "Historical Background and Notes on the Development of Electro-Motive," 20; Hamilton, digest of remarks at the EMD Silver Anniversary Dinner, Chicago, October 24, 1947, GMI, folder 76–16.2.

63. Richard M. Dilworth to F. K. Fildes, November 8, 1929; Dilworth to Fildes, February 13, 1930, both in PRR Collection, box 477, folder 415.3/15.

Chapter III
First-Mover Advantages and the Decentralized Corporation

1. See also: Albert Churella, "Corporate Response to Technological Change: The Electro-Motive Division of General Motors During the 1930s," paper presented at a meeting of the Economic and Business Historical Society, Nashville, Tennessee, April 24, 1993; and Churella, "Corporate Culture and Marketing in the American Railway Locomotive Industry: American Locomotive and Electro-Motive Respond to Dieselization," *Business History Review* 69 (Summer 1995): 191–229.

2. *United States of America vs. General Motors Corporation*, United States District Court, Southern District of New York, Case #61CR356, filed April 12, 1961, National Archives, Northeast Region, 3; Remarks of Charles F. Kettering at lun-

cheon given by Alfred P. Sloan Jr., New York, October 28, 1937, AAR; Charles F. Kettering speech, EMD silver anniversary, Chicago, October 24, 1947, GMI, folder 76–16.2.

3. Borland, "Research Report," GMI, 7, 35, 39.

4. In his book *Boss Kettering* (New York: Columbia University Press, 1983), Stuart W. Leslie provides an excellent in-depth study of the life and career of Charles F. Kettering. Only a small portion of this book (pp. 267–73) is devoted to diesel locomotives, an indication of the breadth of Kettering's interests and abilities.

5. For additional information on Kettering and tetraethyl lead, see Alan P. Loeb, "Birth of the Kettering Doctrine: Fordism, Sloanism and the Discovery of Tetraethyl Lead," *Business and Economic History* 24:1 (Fall 1995): 72–87.

6. T. A. Boyd, provisional draft, "Advances in Engines and Fuels: A History of Vital Pioneering in the Field," 1958, GMI, folder 18/3, 71.

7. C. F. Kettering to Carl G. Fisher, July 24, 1928, reprinted in "Research Report TI-8," 11.

8. "Research Report TI-8," 17, 37. Nelson C. Dezendorf, address at the annual dinner of the Traffic Club of St. Louis, St. Louis, January 15, 1947, AAR; Remarks by C. F. Kettering at a luncheon given by Alfred P. Sloan Jr., New York, October 28, 1937; *Fortune* 19 (March 1939), 49.

9. "Research Report TI-8," 289–90, *Fortune* 19 (March 1939), 51.

10. Boyd, "Advances in Engines and Fuels," 73–80; Charles Kettering to Alfred Sloan, December 18, 1928, reprinted in "Research Report TI-8," 20; same volume, 51.

11. In describing the 1953 merger of pharmaceutical producers Merck and Sharp & Dohme, Louis Galambos concludes that the blending of the cultures of these two companies initiated an "innovation cycle" that lasted until the mid-1970s. This type of creative blending of complementary corporate cultures also occurred at EMC during the 1930s and produced an "innovation cycle" that lasted until the early 1980s. Galambos, "The Innovative Organization: Viewed from the Shoulders of Schumpeter, Chandler, Lazonick, et al.," *Business and Economic History* 22:1 (Fall 1993): 79–91.

12. "Research Report TI-8," 57, 65, 200; Reck, *On Time*, 67, 118.

13. Early fuel injectors typically required fuel-line pressures of six or seven thousand pounds per square inch. "Statement of Harold L. Hamilton," Senate Hearings, November 10, 1955, 2419.

14. Clyde W. Truxell interview, The Kettering Archives, 1965 Oral History Project, March 10, 1961, 3–5 (quote), W. F. Kiesel to F. W. Hankins, July 12, 1934, PRR Collection, box 598, file 4.

15. Salisbury applied for a patent on his version of the unit fuel injector on April 16, 1928. Winton was not willing to install unit injectors on its engines until February 1932, however. Harold L. Hamilton, "Historical Background and Notes on the Development of Electro-Motive," November 22, 1946, GMI, folder 76–16.1, 25–6; "Research Report TI-8," 37, 291; *Automotive Industries* 76 (May 1, 1937), 640; *Fortune* 19 (March 1939), 51; *Scientific American* 221:2 (August 1969), 108–17.

16. Harlow H. Curtice, "Material on Antitrust and Related Problems," excerpts from hearings of the Senate Subcommittee on Antitrust and Monopoly Legislation, December 9, 1955, GMI, folder 83–4.5, 65 (quote).

17. *Baldwin Locomotives* 15:3 (January 1937), 24–27; *Railway Age* 96:19 (May 12, 1934), 685–89; 146:4 (January 26, 1959), 14; Garmany, *Southern Pacific Dieselization*, 16.

18. J. B. Hill to Lyman Delano, August 24, 1939, L & N Collection, box 1, folder A-15113; "Research Report TI-8," 276; Boyd, "Advances in Engines and Fuels," 76–78; "Statement of Harold L. Hamilton," Senate Hearings, November 10, 1955, 2437.

19. The Model 201 designation, like that of all subsequent GM diesel locomotive engines, was based on the cubic-inch displacement of each cylinder.

20. GM staged the first tests of a single-cylinder test engine, built in cooperation with Winton, in May 1931. The Model 201 eventually evolved into the smaller Model 546, GM's first successful truck engine. "Research Report TI-8," 80–84, 139, 148, 169, 291–92. Clyde W. Truxell interview, The Kettering Archives, 1965 Oral History Project, March 10, 1961, GMI. An untitled history of the Cleveland Diesel Division (the corporate successor to Winton), ca. 1962, describes "close cooperation and study" between GM and Navy officials between 1933 and 1940, including, in 1934, the establishment of "the first Navy Training School for Diesel Specialists" at Cleveland Diesel (Winton). (Untitled history, 11–12, GMI, folder 76–16.1.)

21. "Research Report TI-8," 254.

22. Charles Kettering to John Pratt, July 15, 1935, "Research Report TI-8," 235, 164, 181, 193.

23. "Statement of Harold L. Hamilton," Senate Hearings, November 10, 1955, 2421–22; Hamilton, "Historical Background and Notes on the Development of Electro-Motive"; Hamilton, interview by members of the GM Research Laboratories, October 14, 1957, in "Research Report TI-8," 90 (quote). Hamilton was not forced from power following the GM takeover, as is so often the case when large companies take over smaller ones. Hamilton remained president of GM's new Electro-Motive subsidiary; and, when that company was later absorbed into GM's corporate structure as the Electro-Motive Division, Hamilton was elevated to the status of a GM vice president. Throughout his career with Electro-Motive, both as an independent company and as an affiliate of GM, Hamilton remained a strong advocate for the use of diesel power in railroad locomotives—a stronger advocate, in fact, than more senior GM executives such as Sloan. To a certain extent Hamilton's career thus paralleled that of Sloan himself. Sloan joined the Hyatt Roller Bearing Company in 1895 and, during the next twenty years, helped the company to become an important supplier to the growing automobile industry. In 1916, William Durant, president of General Motors, purchased Hyatt and, recognizing Sloan's business acumen, invited him to serve as a GM vice president. After Durant lost control of GM to the DuPont interests, Sloan rose still higher in the GM organization. (Alfred P. Sloan Jr., *My Years with General Motors* [New York: Doubleday, 1964], 17–26.)

24. Hamilton, "Historical Background and Notes on the Development of Electro-Motive," 25.

25. "Statement of Harold L. Hamilton," Senate Hearings, November 10, 1955, 2434. EMC continued to lose money throughout the 1930s. GM subsidized EMC by funneling salaries for its executives through the Chevrolet Division.

26. For a more thorough discussion of the role of a community of technical practitioners, see Constant, "The Social Locus of Technological Practice."

27. "Statement of Harold L. Hamilton," Senate Hearings, November 10, 1955, 2422.

28. Ibid., 2435.

29. Ibid., 2433.

30. "Research Report TI-8," 291.

31. "Statement of Harold L. Hamilton," Senate Hearings, November 10, 1955, 2439.

32. Ibid., 174, 365–72 (quote); C. F. Huddle interview, The Kettering Archives, 1965 Oral History Project, November 10, 1964, GMI.

33. Sloan, *My Years with General Motors*, 349. GM scrapped both engines immediately after the fair closed.

34. *Fortune* 38 (July 1948), 76–81, 144–49; Ralph Budd, "Light-Weight Diesel-Electric Trains," *Civil Engineering* 8:9 (September 1938), 592–95; Budd, remarks at EMD's silver anniversary dinner, Chicago, October 24, 1947, GMI, folder 76–16.2; Coverdale and Colpitts, Consulting Engineers, "Report on Light-Weight Trains of the Zephyr Type," January 15, 1935, AAR; *General Electric Review* 40:9 (September 1937), 421–28.

35. Cyrus R. Osborn interview, The Kettering Archives, 1965 Oral History Project, June 9, 1964, GMI.

36. Nelson C. Dezendorf interview, The Kettering Archives, 1965 Oral History Project, April 6, 1961, GMI.

37. Speech by Ralph Budd at EMD's Silver Anniversary Dinner, Chicago, October 24, 1947, GMI, folder 76–16.2.

38. For an excellent account of the *Zephyr*'s construction and of the career of its designer, Edward Budd, see: Mark Reutter, "The Life of Edward Budd, Part 1: Pulleys, McKeen Cars, and the Origins of the *Zephyr*," *Railroad History* 172 (Spring 1995): 5–34, and Reutter, "The Life of Edward Budd, Part 2: Frustration and Acclaim," *Railroad History* 173 (Autumn 1995): 58–101. Ralph Budd and Edward Budd were apparently very distantly related, but they had not met prior to the design of the *Zephyr*.

39. "Research Report TI-8," 200; Harlow H. Curtice, speech delivered at the American Institute of Consulting Engineers, New York, November 27, 1956, GMI, folder 83–1.9–60. GM engineers probably meant marker lamps when they described the missing "tail lights"—no doubt they were more familiar with automotive jargon than that of the railroads.

40. This movie should not be confused with the more successful 20th Century Fox film *Silver Streak* (1976), which starred Gene Wilder, Jill Clayburgh, and Richard Pryor. John Walker, ed., *Halliwell's Film Guide* (London: Harper Perennial, 1995), 977; Berge and Loftus, *Diesel Motor Trains*, 9–10; Chicago, Burlington, and Quincy Railroad, response to Reconstruction Finance Corporation questionnaire, March 5, 1937, The Records of the Reconstruction Finance Corporation, The National Archives, Washington (Record Group 234), box 188.

41. Kirkland, *Dawn of the Diesel Age*, 83.

42. *Railway Age* 94:21 (May 27, 1933), 761–62; 96:5 (February 3, 1934), 184–96; 97:15 (October 13, 1934), 427–30; 98:6 (February 9, 1935), 220–28; 98:23 (June 8, 1935), 875–79; 100:22 (May 30, 1936), 864–75; Berge and Loftus, *Diesel Motor Trains*, 9; Garmany, *Southern Pacific Dieselization*, 94.

43. John H. White Jr., *The American Railroad Passenger Car* (Baltimore: Johns Hopkins University Press, 1978), 167–81.

44. *General Electric Review* 40:9 (September 1937), 421–28; Berge and Loftus, *Diesel Motor Trains*, 21; Garmany, *Southern Pacific Dieselization*, 105.

45. Reck, *The Dilworth Story*, 49.

46. R. K. Evans to Alfred Sloan, April 10, 1937, reprinted in "Research Report TI-8," 259; S. M. DuBrul, et al., excerpts from transcript of interview of Charles F. Kettering concerning the history of the GM diesel engine and locomotive, July 18, 1957, GMI, folder B3/17 (quote).

47. Address by Charles F. Kettering before the employees of the Electromotive Division of General Motors, Chicago, June 13, 1950, GMI, folder 10/28 (quote); Sloan, *My Years with General Motors*, 350.

48. *Railway Age* 101:19 (November 7, 1936), 696; 102:23 (June 5, 1937), 960; 105:19 (November 5, 1938), 680; Boyd, "Advances in Engines and Fuels," 85; GM-EMD, *The Diesel Locomotive: Preface of a New Era*, ca. 1951, GMI, folder 83–12.101, 16; Reck, *On Time*, 90–94, 120–21.

49. "Research Report TI-8"; W. F. Kiesel to F. W. Hankins, August 13, 1935, PRR Collection, box 597, folder 416/9.

50. Winton itself followed a path that was increasingly divergent from that of EMC. The Electro-Motive Company and the Winton Engine Corporation had been consolidated under the latter name on January 1, 1933. In November 1934, it became the Winton Division of GM, and then the Cleveland Diesel Engine Division in January 1938. At the same time, GM also opened the first facilities of the new Detroit Diesel Engine Division, established in April 1937. Simultaneously with its introduction of the Model 567 locomotive engine, GM announced plans to mass produce a wide range of smaller two-cycle diesel engines at Detroit and Cleveland. GM also established consolidated testing facilities in Detroit to accommodate Cleveland, Detroit, and La Grange. While EMC would market engines of six hundred or more horsepower, Cleveland Diesel would sell the smaller engines through existing auto parts distributors. *Automotive Industries* 76 (May 1, 1937), 640; 78 (January 22, 1938), 95, 101; *Railway Age* 94:1 (January 7, 1933), 30; 104:5 (January 29, 1938), 232; Untitled history of the Cleveland Diesel Division, ca. 1962, GMI, folder 76–16.1, 4,8; "Research Report TI-8," 293.

51. "Statement of Harold L. Hamilton," Senate Hearings, November 10, 1955, 2442.

52. *Railway Age* 99:14 (September 14, 1935), 344; 99:19 (November 9, 1935), 595–97, 611; 99:21 (November 23, 1935), 684; 102:21 (May 22, 1937), 855–64; 102:22 (May 29, 1937), 919; 103:25 (December 18, 1937), 868–70; "Research Report TI-8," 206; Kirkland, *Dawn of the Diesel Age*, 166–71; Reck, *The Dilworth Story*, 58; Reck, *On Time*, 98–99. Blame for the Santa Fe locomotive fire actually lay with railroad service crews, who improperly filled the locomotive's fuel tanks. Wallace Abbey, personal correspondence, November 26, 1995.

53. "Statement of Harold L. Hamilton," Senate Hearings, November 11, 1955, 2450–51 (quotes); GM exhibits, Senate Hearings, December 9, 1955, 3963.

54. "Research Report TI-8," 233, 247, 254, 292; R. K. Evans to Alfred Sloan, April 10, 1937, reprinted in "Research Report TI-8," 259 (quote); Reck, *On Time*, 118–19.

55. EMC's decision to integrate production by manufacturing its own electrical equipment may have been the factor that caused GE, in an attempt to secure a new market for its products, to sign a joint-production agreement with ALCo in 1940—an agreement that restricted the latter company's ability to develop its own organizational capabilities.

56. *Steel* 97 (August 5, 1935), 16.

57. Burnham Finney, "Where Diesels Are Railroaded," *American Machinist* 80 (April 22, 1936): 331–34.

58. *Railway Age* 96:19 (May 12, 1934), 685–89; GM-EMD, *The Diesel Locomotive: Preface of a New Era*, 16; Nelson C. Dezendorf, speech before the New York Society of Security Analysts, New York, June 1, 1951, AAR.

59. *Railway Age* 106:7 (February 18, 1939), 303; 107:20 (November 11, 1939), 736; 108:11 (March 16, 1940), 497; 110:11 (March 15, 1941), 494; Jerry A. Pinkepank, *The Second Diesel Spotter's Guide* (Milwaukee: Kalmbach Publishing, 1973), 118–24.

60. *Wall Street Journal*, October 28, 1938; Garmany, *Southern Pacific Dieselization*, 110.

61. T. T. Blickle, "The History, Development, Operation, and Performance of Diesel Locomotives on the Santa Fe Railway," paper presented at the VIII Pan-American Railway Congress, Washington, D.C., June 12, 1953, Atchison, Topeka & Santa Fe Railway Collection, Kansas State Historical Society, Topeka (hereafter referred to as the ATSF Collection).

62. *Railway Age* 108:3 (January 20, 1940), 159; 109:11 (September 14, 1940), 378–79; 110:6 (February 8, 1941), 287–88; 110:11 (March 15, 1941), 452–58; *Railway Locomotives and Cars* 127 (September 1953), 84–89; *Wall Street Journal*, March 5, 1941; GM-EMD, "Conference Leader's Outline," Subject III, Unit 2A, GMI, folder 76–1.60, 28.

63. Reck, *The Dilworth Story*, 71–72.

64. "Statement of Cyrus R. Osborn," Senate Hearings, December 9, 1955, 3950.

65. A typical comment regarding locomotive design came from F. W. Hankins, the chief of motive power for the Pennsylvania Railroad—"I suggest that, from a general appearance standpoint, you arrange, through your General Motors affiliation, to have your streamlining Engineers [sic] look over this job, with the possibility of producing the same engine with a better outside appearance." F. W. Hankins to Paul R. Turner, July 30, 1934, PRR Collection, box 598, folder 4.

66. Reck, *The Dilworth Story*, 50.

67. Some early EMC switchers did use cast frames, although this was not common practice.

68. One 5,400-hp FT diesel contained 16.8 miles of welded joints. *Railway Age* 95:11 (September 9, 1933), 363–66; *Wall Street Journal*, October 28, 1938; Hamilton, "The Development of the Diesel Locomotive," address to the Pacific Railway Club, Los Angeles, June 9, 1949, reprinted in *Proceedings: The Journal of the Pacific Railway Club* 33:2 (May 1949), 5–22; Remarks by B. B. Brownell at the EMD 50th anniversary luncheon, September 8, 1972, GMI, folder 76–16.3; Hamilton, "Historical Background and Notes on the Development of Electro-Motive," 15; GM-EMD, *The Diesel Locomotive: Preface of a New Era*, 16; Untitled history of the Cleveland

Diesel Division, ca. 1962, GMI, folder 76–16.1; GM-EMD, "Conference Leader's Outline," Subject III, Unit 3C, GMI, folder 76–1.60a.

69. *Fortune* 19 (March 1939), 138; 38 (July 1948), 76–81, 144–49; "Statement of Cyrus R. Osborn," Senate Hearings, December 9, 1955, 3950.

70. *Railway Age* 105:5 (July 30, 1938), 203–4.

71. For an in-depth treatment of the effects of social and cultural differences on technological development, see Hughes, *Networks of Power*, 175–261.

72. Although Hamilton indicated that GMAC only offered financing for two or three years during the depth of the depression, GMAC actually continued to provide this service until at least 1955. In that year, for example, the PRR purchased some two hundred GP-9 locomotives, with financing arranged through GMAC. "Statement of Harold L. Hamilton," Senate Hearings, November 10, 1955, 2458–60 (quote).

73. *Railway Age* 99:6 (August 10, 1935), 193; *Wall Street Journal*, September 1, 1938, 1, 4; O. F. Brookmeyer to J. B. Hill, January 23, 1939; Fitzgerald Hall to J. B. Hill, July 22, 1939; both in L & N Collection, box 1, folder A-15113.

74. Remarks of Ralph Budd at a luncheon given by Alfred P. Sloan Jr., New York, October 28, 1937, AAR.

75. C. H. Burgess to Robert S. Macfarlane, May 13, 1954; EMD, "Diesel Locomotive School," 1945, both in NP Collection), box 494, folder 1214–31, 137.E.2.4.(F); *Railway Age* 102:20 (May 15, 1937), 832–33; 117:16 (October 14, 1944), 600; 118:12 (March 24, 1945), 541–42; 130:14 (April 9, 1951), 41–44; Untitled history of the Cleveland Diesel Division, 11–12; Reck, *On Time*, 169–70.

76. *Railway Age* 130:14 (April 9, 1951), 41–44; EMD, "The Electro-Motive Commitment to Service," ca. 1970, GMI, folder 76–16.4.

77. *Railway Age* 117:19 (November 4, 1944), 683–86; 117:20 (November 11, 1944), 722–23; "Statement of Cyrus R. Osborn," Senate Hearings, December 9, 1955, 3957 (quote); Reck, *On Time*, 169–70.

78. Lyman Delano to J. B. Hill, July 8, 1941, L & N Collection, box 57, folder 1883-A.

79. Another source puts the amount at $4 million. New York *Journal-American*, April 18, 1961, in GMI, folder 11/133.

80. The higher figure is from "Statement of Harold L. Hamilton," Senate Hearings, November 10, 1955, 2451. The $22 million figure may be somewhat high, since Hamilton himself admitted that even he could not accurately estimate what portion of GM's diesel engine R and D budget went toward locomotive development. *Fortune* 19 (March 1939), 46, 51; 38 (July 1948), 76–78, 144–49; "Research Report TI-8," 276–77; Hamilton, "The Development of the Diesel Locomotive"; Reck, *The Dilworth Story*, 63; Reck, *On Time*, 119.

81. Charles E. Wilson, speech at EMD silver anniversary dinner, Chicago, October 24, 1947, GMI, folder 76–16.2.

82. *Iron Age* 140 (November 11, 1937), 85–89; *Railway Age* 105:19 (November 5, 1938), 680; Harlow H. Curtice, speech delivered at the American Institute of Consulting Engineers, New York, November 27, 1956, GMI, folder 83–1.9–60.

83. Reck, *On Time*, 110.

84. "Research Report TI-8," 200.

85. See Chandler, *Strategy and Structure*, for example.

Chapter IV
ALCo and Baldwin

1. For a broader analysis of the reluctance of established companies to embrace new technologies and the ability of new entrants to adopt this technology without jeopardizing existing product lines or organizational strengths, see Richard Foster, *Innovation: The Attacker's Advantage*.

2. American Locomotive Company, "Growing with Schenectady," 1948, AAR, 9–12; William S. Frame, *History of the Schenectady Plant of the American Locomotive Company*, first presented at the Schenectady County Historical Society, Schenectady, New York, May 8, 1945, AAR, 4–6, 10.

3. Report of Appraisers, Supreme Court of the State of New York, in the matter of the applications of Samuel Posen, Clara Berdon, Belle Kuller, Edgar Scott, et al., 1944, AAR; "Manufacturing Capabilities, Alco Products, Inc., Latrobe, Pa.," 1963(?), ALCo Collection, box 11; *Commercial and Financial Chronicle*, October 29, 1945, 2010; *Railway Age* 103:14 (October 2, 1937), 468–69; 105:25 (December 17, 1938), 889, 897; 106:14 (April 8, 1939), 627–28.

4. By 1935, diesel switchers were 20 percent of ALCo's locomotive output. Dana T. Hughes, "A History of Schenectady Operations of Alco Products, Inc.," reprinted from the Schenectady *Union-Star*, April 22, 1955, ALCo Collection, box 11; "New Muscle in Diesels," reprinted from *Alco Review* Spring-Summer, 1959, ALCo Collection, box 1; *Alco Products Review*, 5:2–3 (Spring/Summer 1956), 15; American Locomotive Company, "Growing with Schenectady," 1948, AAR, 35; American Locomotive Company, "Light Locomotive Parts and General Products Catalog, #10057," no date, ACHS, box 897.

5. Between 1901 and 1910, ALCo sold an average of two thousand locomotives per year. *Railway Age* 92:1 (January 2, 1932), 43–44; 94:5 (February, 4, 1933), 172; 96:4 (January 27, 1934), 156; 98:4 (January 26, 1935), 157; 100:1 (January 4, 1936), 72–73.

6. *The Magazine of Wall Street*, 60:3 (May 22, 1937), 152–55; 77:8 (January 19, 1946), 459; "Report of Appraisers, Supreme Court of the State of New York, in the matter of the Applications of Samuel Posen, Clara Berdon, Belle Kuller, Edgar Scott, et al., 1944, AAR, 22.

7. Marx, "Technological Change and the Theory of the Firm," 9; *Alco Products Review* 5:2–3 (Spring/Summer 1956), 15.

8. *Diesel Railway Traction* 15:348 (May 1961), 191–97; *Railway Age* 108:24 (June 15, 1940), 1069; Garmany, *Southern Pacific Dieselization*, 182.

9. By the mid-1930s ALCo operated plants at Schenectady (steam and diesel locomotives); Dunkirk, New York (Alco Products Division); Richmond, Virginia (steam locomotive parts and accessories); Auburn, New York (diesel engines); Latrobe, Pennsylvania, and Chicago Heights, Illinois (Railway Steel Spring Division). *The Magazine of Wall Street*, 60:3 (May 22, 1937), 152–55, 192–93; *Barron's*, June 7, 1937, 9; *Railway Age*, 80:27 (June 5, 1926), 1501–2; 105:25 (December 17, 1938), 889, 897; 108:23 (June 8, 1940), 1011; *Business Week*, June 8, 1946, 66–68.

10. *Barron's*, June 7, 1937, 9.

11. *Barron's*, October 19, 1942, 20; *The Magazine of Wall Street*, 77:8 (January 19, 1946), 460.

12. *Railway Age*, 108:10 (March 9, 1940), 445, 452.

13. ALCo, 1940 annual report, 5; *Railway Age*, 108:10 (March 9, 1940), 445, 452 (quote); 120:2 (January 12, 1946), 146; 120:18 (May 4, 1946), 906.

14. *Railway Age*, 104:19 (May 7, 1938), 796–801.

15. William C. Dickerman, address to the Western Railway Club, April 25, 1938, quoted in "Senate Hearings," December 9, 1955, 3955.

16. ALCo, 1954 annual report, 14; *New York Times*, February 26, 1954, 32; *Railway Age*, 120:2 (January 12, 1946), 146.

17. *Railway Age*, 116:7 (February 12, 1944), 369.

18. Brown, *The Baldwin Locomotive Works*, 216.

19. The Baldwin Locomotive Works, *The Story of Eddystone* (Philadelphia: The Baldwin Locomotive Works, 1929); *Baldwin Locomotives* 9:4 (April 1931), 3–15; 10:3 (January 1932), 21–34; *System* 38 (November 1920), 815–18, 926–32; Samuel M. Vauclain, *Mass Production within One Lifetime* (Princeton: Princeton University Press, 1937).

20. Brown, *The Baldwin Locomotive Works*, 174.

21. Ibid., 189–94.

22. *Baldwin Locomotives* 9:4 (April 1931), 3–15; *Railway Age* 105:27 (December 31, 1938), 951, 954; 108:6 (February 10, 1934), 282–83.

23. Brown, *The Baldwin Locomotive Works*, 81–85.

24. Whitcomb began as a producer of coal mining machinery in 1878. Baldwin acquired a majority interest in the company, and made it a subsidiary in 1931. On May 31, 1940, Baldwin purchased the last 8.05 percent of the outstanding stock of the Whitcomb Locomotive Company, thus making it a wholly owned subsidiary. *Baldwin Locomotives* 12:3 (January 1934), 22–23 (quote); 3 & 4 (February 1940), 5–8.

25. In September 1939, Baldwin transferred the sales and engineering staff of the Whitcomb Locomotive Company from Eddystone to Rochelle, Illinois. These employees, who were rapidly gaining experience in the production and marketing of ever larger gasoline locomotives, would have been immensely valuable to Baldwin's diesel locomotive program. *Baldwin Locomotives* 18:1 & 2 (October 1939), 10; 18:3 & 4 (February 1940), 5–8; Baldwin Locomotive Works, 1939 and 1940 annual reports; Kirkland, *Dawn of the Diesel Age*, 64.

26. Baldwin Locomotive Works, 1936 annual report; *Moody's Analyses of Industrial Investments, 1920* (New York: Moody's Investor's Service, 1920), 1319; *Moody's Manual of Industrial Securities, 1926* (New York: Moody's, 1926), 478; *Moody's Manual of Industrial Securities, 1931* (New York: Moody's, 1931), 990. Lima's dividend policy was somewhat more restrained than that of its larger competitors.

27. Baldwin had facilities at Eddystone (Philadelphia), Pennsylvania; Nicetown, Pennsylvania (Midvale); Burnham, Pennsylvania (Standard Steel); Granite City, Illinois (General Steel Castings); and Rochelle, Illinois (Whitcomb). *Baldwin Locomotives*, 1:4 (April 1923), 24–26; 9:4 (April 1931), 3–15; 20:1 (June 1942), 33; *The Magazine of Wall Street*, 60:3 (May 22, 1937), 152–55, 192–93; *Barron's*, June 7, 1937, 9; *Railway Age*, 80:27 (June 5, 1926), 1501–2; 105:25 (December 17, 1938), 889, 897; 108:23 (June 8, 1940), 1011; *Business Week*, June 8, 1946, 66–68; John Bonds Garmany, *Southern Pacific Dieselization* (Edmonds, Washington: Pacific Fast Mail, 1985), 73.

28. Brown, *The Baldwin Locomotive Works*, 172–83.

29. *Baldwin Locomotives* 8:1 (July 1929), 3; *Railway Age* 86:14 (April 6, 1929), 791–92; 108:6 (February 10, 1940), 282–83; Marvin W. Smith, *Samuel Vauclain: Courageous Pioneer, Believer in America!* (New York: The Newcomen Society in North America, 1952); Vauclain, *Mass Production within One Lifetime*.

30. *Baldwin Locomotives* 5:1 (July 1926), 43–49.

31. *Railway Age* 88:25D (June 25, 1930), 1548D140–44. After World War II, Baldwin and several other companies attempted to develop steam turbine locomotives. These were failures, however, since steam turbine technology proved better suited for larger applications such as ships and electrical generating facilities.

32. *Railway Age* 108:6 (February 10, 1934), 282–83; Vauclain, *Mass Production within One Lifetime*, 22 (quote).

33. *Railway Age* 92:25, (June 18, 1932), 1009–10; *Baldwin Locomotives* 13:2 (October 1934), 12–17.

34. *Baldwin Locomotives* 14:1 (April–July 1935), 11–20.

35. The parallel in the auto industry is of course Henry Ford, who so thoroughly purged his company of dissent that it was unable to respond to changing customer demands during the 1920s.

36. *Baldwin Locomotives* 5:1 (July 1926), 43–49; 18:3 + 4 (February 1940), 31–33; *Railway Age* 79:15 (October 10, 1925), 645–48; Garmany, *Southern Pacific Dieselization*, 73; Kirkland, *Dawn of the Diesel Age*, 85–86.

37. Pennsylvania Railroad tests of these locomotives uncovered such difficulties as a malfunctioning battery-charging system; an overheated engine (the result of rubber scraps in the cooling system); excessive smoke; oil leaks; a broken air compressor coupling; starting difficulties caused by cold weather; excessive wear on the engine cylinders; unspecified electrical problems; stalling; and a history of frequent crankcase explosions. "Tests of Baldwin Diesel Electric Switching Engine . . . ," February and March 1940; PRR Collection, box 597, folder 416/7.

38. Appraisal of Baldwin diesel locomotives 58501 and 61000, ca. 1942–1943, PRR Collection, box 537, folder 410.043.1/16 (quotes).

39. Samuel Vauclain to Fred W. Hankins, September 1929, PRR Collection, box 518, folder 400.3/18.

40. C. L. Cummins to J. T. Wallis, January 4, 1928, PRR Collection, box 334, folder 416/15.

41. Samuel M. Curwen to J. T. Wallis, January 7, 1928, PRR Collection, box 334, folder 416/15.

42. *The Commercial and Financial Chronicle*, March 9, 1935, 1651; June 22, 1935, 4226; August 10, 1935, 908–9; *Railway Age* 98:9 (March 2, 1935), 346; "Special Master's Report on Plan of Reorganization," December 2, 1935; Baldwin Locomotive Works, "To the Bondholders and Stockholders of the Baldwin Locomotive Works," March 24, 1936, AAR.

43. *Business Week*, June 8, 1946, 66–68.

44. Baldwin Locomotive Works, 1938 and 1939 annual reports.

45. The Fisher brothers purchased a block of 120,000 shares of Baldwin stock in 1927, largely to have some measure of control over the production of large hydraulic sheet metal stamping presses at the Southwark Company. *Business Week*, June 8,

1946, 66–68; *Railway Age* 96:11 (March 17, 1934), 388; 104:11 (March 12, 1938), 475–76; 105:10 (September 3, 1938), 354–55; Baldwin Locomotive Works, 1938 annual report.

46. *Baldwin Locomotives* 17:3 & 4 (February 1939), 3; 20:3 (October 1943), 8; *Railway Age* 105:27 (December 31, 1938), 951.

47. *Baldwin Locomotives* 10:3 (January 1932), 21–34.

48. G. H. Woodroffe to J. G. Earles, December 4, 1939, in "Correspondence: VO Engine," Baldwin Locomotive Works Collection, Pennsylvania State Archives, Harrisburg (RG 427, hereafter referred to as the Baldwin Collection), box C-12–3, carton 3 (quote); Baldwin diesel locomotive specifications, Baldwin Collection, box 1027.

49. *Baldwin Locomotives* 15:2 (October 1936), 3–8; *Railway Age* 101:21 (November 21, 1936), 746–48; Baldwin Locomotive Works, 1935 annual report; GM-EMD, "Conference Leader's Outline," Subject III, Unit 1B, 39, 43–4, GMI; Kirkland, *Dawn of the Diesel Age*, 43–44, 151.

50. B. G. Mellin to E. J. Harley, August 30, 1946, in "Planning and Simplification Committee," Baldwin Collection, box C-12–5, carton 6.

51. A. T. Brenser to Lewis W. Metzger, April 13, 1945; John F. Kirkland to G. L. Bader, September 13, 1945, both in "Correspondence: VO Engine," Baldwin Collection, box C-12–3, carton 3.

52. *Baldwin Locomotives* 17:2 (October 1938), 17–9; 18:3 & 4 (February 1940), 31–33; *Railway Age* 107:24 (December 9, 1939), 906; 107:27 (December 30, 1939), 992–95, 998; *Railway Mechanical Engineer* (February 1940), 56–60, 64; Kirkland, *Dawn of the Diesel Age*, 187.

53. *Railway Age* 102:1 (January 2, 1937), 72; 104:1 (January 1, 1938), 78–79; 106:1 (January 7, 1939), 81; 108:1 (January 6, 1940), 81.

54. Max Essl to G. H. Woodroffe, January 20, 1940, in "Correspondence: VO Engine," Baldwin Collection, box C-12–3, carton 3.

55. R. B. Crean, memorandum, November 6, 1946, in "Planning and Simplification Committee," Baldwin Collection, box C-12–5, carton 6.

56. *Baldwin Locomotives* 18:3 & 4 (February 1940), 5–8.

57. "Statement by O. DeGray Vanderbilt III," Senate Hearings, November 9, 1955, 2362, emphasis added.

58. W. E. Smith to J. B. Hill, August 23, 1939, Louisville and Nashville Railroad Collection at the University of Louisville (record group 123, hereafter referred to as the L & N Collection), box 1, folder A-15113.

59. ALCo and Baldwin thus support Kenneth Lipartito's assertion that "firm routines are even harder to change than we imagined." Lipartito, "Culture and the Practice of Business History," 29.

60. By the 1930s, the steam locomotive builders could find a precedent to justify their skepticism of diesel locomotive technology. Elaborate promises by the proponents of widespread mainline electrification in the early twentieth century had not come to fruition, and perhaps the diesel locomotive would eventually suffer the same fate. Executives understood that if they tied their careers to the diesel locomotive and if dieselization became no more widespread than electrification, then their jobs and their companies would be in serious jeopardy.

Chapter V
Policy and Production during World War II

1. The Santa Fe purchased its first diesel locomotive in 1935. It tested Electro-Motive's FT prototype in 1940 and later purchased eighty production-model FT locomotives. By the end of 1952, the Santa Fe owned 1,410 diesels from various builders. John B. McCall, "Dieselisation of American Railroads: A Case Study," *The Journal of Transport History* 6:2 (September 1985), 1–17; Blickle, "The History, Development, Operation, and Performance of Diesel Locomotives on the Santa Fe Railway," ATSF Collection (quote).

2. At that time, there were still some forty thousand steam locomotives in the United States.

3. Most of these new diesel locomotives were assigned to the overburdened western railroads, already struggling to cope with mountainous terrain, heavy traffic, and an arid climate. By 1944, diesels were responsible for 13.5 percent of the freight locomotive mileage on western railroads, a figure that far exceeded that of eastern and southern lines. *Railway Age* 118:20 (May 19, 1945), 888; 122:1 (January 4, 1947), 78; Diesel Engine Manufacturers Association, *Diesel Locomotives*, Chicago, 1945, AAR; Speech by Robert H. Morse Jr., at a meeting of the American Petroleum Institute Coordinating Research Council, November 13, 1946, AAR.

4. Title 32—National Defense, Chapter IX—War Production Board, Subchapter B—Division of Industry Operations, Part 1188—Railroad Equipment, General Limitation Order L-97, April 4, 1942; Andrew Stevenson to C. H. Matthiessen, April 2, 1942 (quote); both in the records of the War Production Board, National Archives, Washington (Record Group 179, hereafter referred to as the WPB Records), box 2162.

5. John Morton Blum, *V Was for Victory: Politics and American Culture during World War II* (New York: Harcourt Brace Jovanovich, 1976), 121–24.

6. Summary of Meeting, Producers of Large Diesel Locomotives Industry Advisory Committee, November 10, 1943, WPB Records, box 2162; *Railway Age* 112:15 (April 11, 1942), 753–54; 110:12 (March 22, 1941), 536; 110:19 (May 10, 1941), 813; 111:4 (July 26, 1941), 170–71; 118:1 (January 6, 1945), 90–94.

7. "Statement of Cyrus R. Osborn," Senate Hearings, December 9, 1955, 3959–60; J. B. Hill to J. J. Pelley, May 18, 1941, L & N Collection, box 1, folder A-15113; Hill to Lyman Delano, November 15, 1941, L & N Collection, box 57, folder 1883-B; Harold L. Hamilton, address to the Pacific Railway Club, Los Angeles, June 9, 1949, reprinted in *Proceedings: The Journal of the Pacific Railway Club* 33:2 (May 1949), 5–22 (quote).

8. J. B. Hill to Joseph B. Eastman, March 10, 1942, L & N Collection, box 57, folder 1883-A.

9. Harold L. Hamilton to J. B. Hill, February 17, 1942, L & N Collection, box 57, folder 1883-A.

10. J. B. Hill to Lyman Delano, July 9, 1943, L & N Collection, box 58, folder 1887.

11. B. W. Scandrett to Charles E. Denney, October 13, 1942, NP Collection, box 507, file 1261–9, 137.E.3.7(B).

NOTES TO CHAPTER V

12. "Summary of Meeting—Builders of Large Steam Locomotives Industry Advisory Committee," December 6, 1944, 7; General Limitation Order L-97, Revocation, July 16, 1945, both in WPB Records; *Railway Age* 119:3 (July 21, 1945), 111; "Statement of Cyrus R. Osborn," Senate Hearings, December 9, 1955, 3959–60.

13. Market share figures are from Thomas G. Marx, "Technological Change and the Theory of the Firm," 9.

14. *Railway Age* 116:17 (April 22, 1944), 778; Diesel Engine Manufacturers Association, *Diesel Locomotives* (Chicago: Diesel Engine Manufacturers Association, 1945), AAR; "Statement of Cyrus R. Osborn," Senate Hearings, December 9, 1955, 3961; Garmany, *Southern Pacific Dieselization*, 189.

15. Blum, *V Was for Victory*, 107–10, 112–16. Geoffrey Perrett, in *Days of Sadness, Years of Triumph: The American People, 1939–1945* (Madison: University of Wisconsin Press, 1985), places less emphasis on the atypical experience of a few highly successful companies.

16. In 1937, Navy officials had asked GM's Research Laboratories to design a compact, lightweight engine suitable for a wide variety of naval applications. GM's efforts resulted in the Model 184 "pancake" engine, so called because the horizontal cylinders were arranged in four tiers of four cylinders each, like a stack of pancakes. This 1,200-hp engine had a then-unheard-of weight of just four pounds per horsepower. The Navy accepted the first Model 184 engine in October 1940, and put it through its first sea trials the following June. EMD ultimately produced 554 pancake engines. The division also supplied 2,100 conventional Model 567 engines to the Navy, primarily for use in landing craft. In all, General Motors sold the Army and Navy 198,000 diesel engines during the war years. *Railway Age* 113:13 (September 26, 1942), 509; "The Development and Growth of General Motors," statement by Harlow H. Curtice before the Subcommittee on Antitrust and Monopoly of the U.S. Senate Committee on the Judiciary, Washington, December 2, 1955, GMI, folder 83–4.2; Borland, "Research Report TI-8," GMI, 293; GM-EMD, *Diesel War Power: The History of Electro-Motive's Diesel Engines in the Service of the United States Navy*, 1945(?), 24–26; Reck, *On Time*, 143–44, 152–53.

17. EMD, "Conference Leader's Outline: Management Conferences," Subject III, Unit 3C, GMI, folder 76–1.60a, 12.

18. *Railway Age* 115:6 (August 7, 1943), 239–40; 116:10 (March 4, 1944), 478; GM-EMD, *Diesel War Power*, 50–51 "Statement of Cyrus R. Osborn," Senate Hearings, December 9, 1955, 3959–62; Reck, *On Time*, 158–59.

19. EMD, "Conference Leader's Outline," Subject III, Unit 3C, 11–12 (quotes).

20. *Business Week*, November 12, 1949, 68–74; *Railway Age* 117:4 (July 22, 1944), 176; EMD, "Conference Leader's Outline," Subject III, Unit 3C, 13, 15, 23, 25–26, 28, 34, 39.

21. EMD, "Conference Leader's Outline," Subject IV, Unit 1 (quote) & Unit 2, GMI, folder 76–1.61.

22. *Railway Age* 115:6 (August 7, 1943), 239–40; 124:4 (June 12, 1948), 1199; 129:8 (August 19, 1950), 43–44; 173:4 (August 28, 1972), 26; GM Executive Bulletin #11, 1949, GMI, folder 83:12.101. One source lists Osborn's prior assignment as the "Alliance" Division, probably a misprint for "Allison."

23. GM's reasons for transferring so many DELCo executives to EMD are unclear, but may have been the result of contacts established by Charles Kettering during his years at DELCo.

24. Jim Marshall, "Giant Jeep," *Collier's*, April 5, 1941, 75.

25. *Railway Age* 132:11 (March 17, 1952), 90–91; Reck, *On Time*, 159–61.

26. EMD, "Conference Leader's Outline," Subject II, Unit 1, GMI, folder 76-1.59.

27. American Locomotive Company, "Growing with Schenectady," 1948, AAR, 32; William S. Frame, "History of the Schenectady Plant of the American Locomotive Company," originally presented at a meeting of the Schenectady County Historical Society, Schenectady, May 8, 1945, AAR, 17–19.

28. ALCo, 1942 annual report, 6; 1943 annual report, 3; 1944 annual report, 3; *Railway Age*, 112:1 (January 3, 1942), 20; 112:17 (April 25, 1942), 835; 118:11 (March 17, 1945), 526; Marx, "Technological Change and the Theory of the Firm," 9.

29. "Title 32—National Defense, Chapter IX—War Production Board, Subchapter B—Division of Industry Operations, Part 1188—Railroad Equipment, General Limitation Order L-97," April 4, 1942; Summary of Meeting, Producers of Large Diesel Locomotives Industry Advisory Committee, November 10, 1943, both in the WPB Records, box 2162.

30. Kirkland, *The Diesel Builders, Vol. 2*, 78–82.

31. Robert B. McColl, Address to the Railroad Executives Conference, Schenectady, New York, March 1948, ALCo Collection, box 6; ALCo 1945 annual report, 10 (quote).

32. *Alco Review*, Spring/Summer, 1959.

33. O. G. Dellacanonica, "Alco's Contribution to the Diesel-Electric Locomotive Industry," reprinted from *The Indian Railway Engineer*, July 1960, ALCo Collection, box 11; "The Alco Engine and Its Development," reprinted from *Diesel Railway Traction*, May 1961, ALCo Collection, box 29; ALCo, *Symposium on Diesel Locomotive Engine Maintenance* (New York: ALCo, 1953), 126.

34. The American Locomotive Company, "The Postwar Outlook for American Locomotive Company: A Memorandum to Shareholders at the War's End," August 23, 1945, AAR.

35. Robert B. McColl, "Diesel-Electric Locomotive Pioneering," luncheon address, St. Louis, March 7, 1947, PRR Collection, box 598, folder 8.

36. *Railway Age* 123:11 (September 13, 1947), 450–53.

37. ALCo, Report of the annual meeting of shareholders, April 19, 1949, ALCo Collection, box 6.

38. *New York Times*, November 10, 1955, 17, 23.

39. EMD, "Conference Leader's Outline," Subject III, Unit 1B, GMI, folder 76-1.60, 32–33; W. E. Smith to J. B. Hill, May 25, 1941 (quote); Smith to Hill, June 21, 1941, both in the L & N Collection, box 1, folder A-15113.

40. J. B. Hill to Lyman Delano, July 9, 1943, L & N Collection, box 58, folder 1887.

41. Hill to F. B. Adams, July 5, 1945, L & N Collection, box 56, folder 1870-B.

42. Garmany, *Southern Pacific Dieselization*, 165.

43. Earl Newsom & Company, Policy Memorandum, ALCo, January 2, 1942, ALCo Collection, box 1.

44. *The Commercial and Financial Chronicle*, February 7, 1942, 593.

45. ALCo won the 1942 Harvard Award "For that advertisement or series of advertisements which contributes to the advancement of advertising as a social force." ALCo, 1942 annual report, 8–9; *Railway Age* 116:17 (April 22, 1944), 793b.

46. ALCo, 1943 annual report, 3; 1944 annual report, 1, 3; *Commercial and Financial Chronicle*, April 2, 1945, 1418; *Railway Age*, 116:22 (May 27, 1944), 1046.

47. In 1941, GE completed a reorganization that made it a diversified, decentralized corporation. The company contained four main operating departments: Appliance and Merchandise, Radio and Television, Lamp, and Apparatus. GE assigned locomotive production to the last division, which, because of its size and diversity, had five vice presidents. At the same time, GE undertook a planning study, completed in 1945, that recommended the "invasion of new business fields with products of G. E. research, [as well as] expansion to maintain . . . relative competitive position" and called for the expenditure of at least $280 million to this end. *Business Week*, November 10, 1945, 44–46.

48. GE press release, October 13, 1943, AAR.

49. *Railway Age* 114:26 (June 26, 1943), 1278–79.

50. As events unfolded, however, few streetcars were built after 1945, and mass transit was never as popular as GE might have wished.

51. In 1942, Baldwin paid dividends on common stock for the first time since 1930. Baldwin Locomotive Works, 1942 annual report.

52. *Railway Age* 116:25 (June 17, 1944), 1186; 118:3 (January 20, 1945), 204 (quote); 118:16 (April 21, 1945), 708–10; *Steel Horizons* 5:3 (1943), 6–7; Baldwin Locomotive Works, 1944 and 1945 annual reports; Baldwin, press release, June 29, 1944, AAR; Garmany, *Southern Pacific Dieselization*, 107.

53. F. B. Adams to J. B. Hill, July 23, 1945, L & N Collection, box 56, folder 1870-B; C. J. Bodemer to E. O. Rollings, January 30, 1943, L & N Collection, box 94, folder 51170, part 1B (quote).

54. J. B. Hill to F. B. Adams, July 25, 1945, L & N Collection, box 56, folder 1870-B.

55. H. H. Sener to Max Essl, July 22, 1941 (quote); Max Essl, "Report of Conference," August 12, 1942; D. R. Staples to G. L. Bader, January 19, 1943; all in "Correspondence: VO Engine," Baldwin Collection, box C-12-3, carton 3; Westinghouse Electric and Manufacturing Company, "Time-Savers Guide for Diesel Electric Maintenance Men," 1942, PRR Collection, box 1417, folder 14; minutes of PRR staff meeting, November 5, 1945, PRR collection, box 326, folder 300.4/40; *Westinghouse-Equipped Diesel-Electric Locomotives—10 to 80 Tons* (East Pittsburgh: Westinghouse Electric and Manufacturing Company, 1942; reprint, Canton, Ohio: Railhead Publications, 1980).

56. *Baldwin Locomotives* 20:1 (June 1942), 10; 20:3 (October 1943), 8; *Railway Age* 114:15 (April 10, 1943), 715; Baldwin Locomotive Works, 1942 and 1943 annual reports.

57. *Business Week*, May 17, 1947, 59–64; *The Magazine of Wall Street*, May 4, 1940, 91–93, 126–27; *Sales Management* 67 (October 1, 1951), 120–24; Fairbanks, Morse & Company, *Pioneers in Industry: The Story of Fairbanks, Morse & Company, 1830–1945* (Chicago: Fairbanks, Morse & Company, 1945), 10, 12–13, 30, 32, 44, 48–49, 60–61; Charles H. Wendel, *Power in the Past, Vol. 2: A History of Fairbanks,*

Morse and Company (Atkins, Iowa: Old Iron Book Company, 1982), 40; Kirkland, *The Diesel Builders, Vol. 1*, 11, 16, 18.

58. "Statement of V. H. Peterson," Senate Hearings, November 9, 1955, 2348.

59. By 1940, Fairbanks-Morse offered a complete line of diesel engines, ranging from 10 to 1,400 hp. *Business Week*, May 17, 1947, 59–64; *Diesel Power and Diesel Transportation*, March 1947, 76–80; *The Magazine of Wall Street*, May 4, 1940, 91–93, 126–27; *Railway Age* 121:4 (July 27, 1946), 126–29; *Steel* 118:2 (January 14, 1946), 73; Fairbanks, Morse & Company, *Pioneers in Industry*, 71–72, 113; Wendel, *Power in the Past*, 91; Kirkland, *The Diesel Builders, Vol. 1*, 22.

60. *Railway Age* 107:12 (September 16, 1939), 399–402; Kirkland, *The Diesel Builders, Vol. 1*, 23–26.

61. Fairbanks, Morse & Company, *Pioneers in Industry*, 72.

62. *Railway Age* 111:18 (November 1, 1941), 724; 116:15 (April 8, 1944), 702; Fairbanks, Morse & Company, *Pioneers in Industry*, 107; Fairbanks-Morse, press release, May 8, 1944, AAR (quote).

63. *Business Week*, May 17, 1947, 59–64.

64. John W. Barriger III, "Super-Power for Super-Railroads," paper presented at the meeting of the Society of Automotive Engineers, June 1948, PRR Collection, box 598, folder 8.

65. *Railway Age* 116:20 (May 13, 1944), 910; 120:16 (April 20, 1946), 840; 123:12 (September 20, 1947), 476–77; Fairbanks, Morse & Company, *Pioneers in Industry*, 134–35; Garmany, *Southern Pacific Dieselization*, 289.

66. *Diesel Power and Diesel Transportation*, March 1947, 76–80; *Railway Age* 116:20 (May 13, 1944), 910; Fairbanks, Morse & Company, *Pioneers in Industry*, 76, 80, 84, 96–97.

67. *Railway Age* 116:20 (May 13, 1944), 910; 117:12 (September 16, 1944), 440–43; Fairbanks, Morse & Company, *Pioneers in Industry*, 72, 145, 155.

Chapter VI
Postwar Dieselization and Industry Shakeout

1. Eastern Railroad Presidents Conference, *A Yearbook of Railroad Information, 1951 Edition* (New York: Eastern Railroad Presidents Conference, 1951). For additional information of the American railroad industry after World War II, see Kent T. Healy, *Performance of the U. S. Railroads since World War II: A Quarter Century of Private Operation* (New York: Vantage Press, 1985).

2. In spite of their commitment to railroad dieselization, EMD executives underestimated the speed with which this transformation took place. For example, EMD predicted that 20 percent of the postwar locomotive demand would be for steam locomotives, with the remainder for diesels.

3. American railroads did order a few additional steam locomotives in the years that followed, but, for all practical purposes, the steam locomotive industry had expired by V-J Day. *Barron's* 28 (October 18, 1948), 29–30; 33 (May 11, 1953), 15–16; *Coal Age* 52:12 (December 1947), 74–78; *Railway Age* 123:20 (November 15, 1947), 829–31; 146:14 (April 6, 1959), 10; 152:2 (January 15, 1962), 16, 103; GM-EMD, "Why America Needs More Diesels Now," 1950, AAR.

4. *Barron's* 29 (January 17, 1949), 4, 46.

NOTES TO CHAPTER VI

5. Once again, EMD's earnings were but a fraction of those of its parent company. While EMD locomotive sales were $85 million in 1960, total GM sales were $12.7 billion. United States of America vs. General Motors Corporation, case # 61CR356, United States District Court, Southern District of New York, filed April 12, 1961, National Archives, Northeast Region, New York, 2, 7–10; Marx, "Technological Change and the Theory of the Firm," 9.

6. Mark Reutter, "The Great Motive Power Struggle: The Pennsylvania Railroad v. General Motors, 1935–1949," *Railroad History* 170 (Spring 1994), 15–33, describes how the PRR replaced older steam loyalists, such as Martin Clement, president; John V. B. Duer, chief electrical engineer; and John F. Deasy, vice president of operations, with younger and more cost-conscious executives. Also useful is Michael Bezilla, "Pennsylvania Railroad Motive Power Strategies, 1920–1950," *Railroad History* 164 (Spring 1991), 43–52.

7. Amy McCallum Bucks to Martin W. Clement, December 23, 1948, PRR Collection, box 598, file 10. Not all members of the general public welcomed the demise of the steam locomotive. Letters addressed to the president of the Northern Pacific included statements such as "Personally I think the 'roads made a big mistake when you started converting to diesel engines.... Nothing on earth could cause the thrill of the old chugging steamers," and "Moreover the deisel [sic] is unfair to engineers, to the firemen, ruins the competence, and does to cause onset of mental disease." R. R. Robinson to Robert S. Macfarlane, January 30, 1956, NP Collection, box 507, file 1261–9, 137.E.3.7(B); Lerois F. Clette Sr., written comment on stock proxy, 1953, NP Collection, box 898, file 2981, 137.G.2.8(F).

8. John F. Deasy to Martin W. Clement, January 6, 1945, PRR Collection, box 653, folder 8.

9. PRR, Minutes of Board Meeting, October 9, 1945, PRR Collection, box 326, folder 300.4/40.

10. Martin W. Clement to W. S. Franklin, November 11, 1946, PRR Collection, box 537, folder 410.043.1/10.

11. PRR, Maryland Division, Chester-Thurlow and York area studies, November 1946, PRR Collection, box 653, folder 5.

12. John F. Deasy to Martin W. Clement, September 30, 11946, PRR Collection, box 326, folder 300.4/38.

13. In one instance, in the coal fields of central Pennsylvania, ten diesels replaced twenty-eight steam locomotives.

14. John F. Deasy to Martin W. Clement, September 22, 1947, PRR Collection, box 654, folder 416/12; Deasy to Clement, January 6, 1945, PRR Collection, box 653, folder 8; J.A.A. to James M. Symes, April 29, 1949, PRR Collection, box 598, folder 10; Symes to H. T. Cover, February 13, 1950, PRR Collection, box 654, folder 416(60B)/5.

15. "Statement of Maintenance and Operation Cost of Diesel Locomotives for the Month of December, 1940, and Seven Months' Cumulative Period ended December 31, 1940," PRR Collection, box 597, folder 416/8; costs are from December 1940.

16. Atchison, Topeka and Santa Fe Railway System Operating Department to R. G. Bennett, October 27, 1941, PRR Collection, box 597, folder 416/6.

17. Memorandum, January 17, 1947, PRR Collection, box 326, folder 300.4/39.

18. "Memo. for Board Meeting," September 15, 1948, PRR Collection, box 653, folder 13; H. T. Cover memorandum, December 17, 1947; "Detentions to Passenger Trains Due to Diesel-Electric Locomotive Failures," January 1948, both in PRR Collection, box 602, folder 416.03/10. The logic behind the PRR's large orders from Baldwin is unclear, given that company's poor products, but likely relates to the proximity of Baldwin's Eddystone and Westinghouse's East Pittsburgh facilities to PRR trackage and to whatever customer loyalty that remained from the days when Baldwin and Westinghouse supplied straight electric locomotives to the PRR.

19. The PRR blamed crankshaft failures for much of the high maintenance costs associated with ALCo diesels. The railroad also complained of defects in the fuel injection system, the turbochargers, the exhaust manifolds, and the piston ring grooves. "Memorandum," May 7, 1957, PRR Collection, box 333, folder 411.043/8.

20. A. J. Greenough to James P. Newell, June 5, 1957; Greenough to Newell, February 20, 1958 (quote); Greenough to Newell, April 25, 1958; "Diesel Electric Locomotive Repair Costs per Mile—By Builder Type—First Eight (8) Months of 1957"; all in PRR Collection, box 333, folder 411.043/8.

21. Memorandum, January 17, 1947, PRR Collection, box 326, folder 300.4/39.

22. Some evidence does indicate that corporate ties influenced the PRR's decision to buy EMD products, since railroad officials were well aware that Pierre S. DuPont served as a director of both the PRR and General Motors. H. W. Jones to John F. Deasy, May 26, 1944, PRR Collection, box 597, folder 416/11.

23. The boom years of diesel locomotive production lasted little more than a decade, however, and as railroads replaced the last of their steam locomotives, orders once again declined. The years between 1957 and 1961 were difficult ones for the entire locomotive industry. In 1950, ten thousand people found work at La Grange, but only six thousand remained in 1961. *Business Week*, September 1, 1956, 52–58; *New York Times*, July 2, 1954, 30; *Railway Age* 135:5 (August 3, 1953), 5; 141:19 (October 29, 1956), 8; *Trains and Travel* 12 (October 1952), 23–30; EMD press release, March 20, 1956, GMI, folder 83–12.99; EMD press release, December 27, 1956, AAR; GM Executive Bulletin #9, 1954, GMI, folder 83–12.101; "Welcome to the Electro-Motive Division of General Motors," ca. 1955, GMI, folder B 3/17; Stenographer's Minutes, United States of America vs. General Motors Corporation, case # 61CR356, May 22, 1961, 9; *Our GM Scrapbook* (Milwaukee: The Kalmbach Publishing Company, 1971).

24. These production figures, while impressive, represent only a small fraction of GM's total output. In 1948 alone, for example, GM's total diesel engine sales were $250 million. *American Machinist*, April 22, 1948, 74–75; *Business Week*, September 1, 1956, 52–58; *Fortune* 38 (July 1948), 76–81, 144–49; *Railway Age* 123:18 (November 1, 1947), 726–28, 739; 128:24 (June 17, 1950), 1201–2; 136:11 (March 15, 1954), 14; 150:16 (April 17, 1961), 10; *Trains and Travel* 12 (October 1952), 23–30; Remarks by Ralph Budd at EMD's Silver Anniversary Dinner, Chicago, October 24, 1947, GMI, folder 76–16.2; Harold L. Hamilton, Proceedings of the Electro-Motive Division, General Motors Corporation, Silver Anniversary Dinner Meeting, Chicago, October 24, 1947, GMI, folder 76–16.2; Digest of Remarks by Harold L. Hamilton, at Silver Anniversary Dinner, Chicago, October 24, 1947, AAR; EMD press release, October 24, 1947, AAR; Nelson C. Dezendorf interview, The Kettering Archives, 1965 Oral History Project, April 6, 1961, 3, GMI.

25. *Railway Age* 121:17 (October 26, 1946), 674–77, 683; 123:20 (November 15, 1947), 877–78; 126:15 (April 9, 1949), 726–30; 127:19 (November 5, 1949), 790–93; 135:19 (November 9, 1953), 28–30. EMD's locomotives, while successful, were far from perfect, and their imperfections created a final opportunity for ALCo to increase its market share. For example, some railroads reported that the GP-7 had a tendency to jackknife when coupled in multiple units. The F-3 also experienced a variety of problems. And, despite substantial production and marketing efforts by EMD, the oddly designed Model BL-2 found few customers in the railroad industry. *Railway Age* 140:3 (January 16, 1956), 6–7; EMD press release, August 3, 1955, AAR; "Have the Expected Economies of Dieselization Been Realized?" report in the *Proceedings and Committee Reports of the Fifty-Sixth Annual Meeting of the American Association of Railroad Superintendents* (Topeka: Hall Lithographing Company, 1952), 212–14; Garmany, *Southern Pacific Dieselization*, 200–201.

26. EMD's voluntary restraint in the diesel locomotive industry was similar to GM's earlier decision to refrain from monopolizing the automobile market. Address by John F. Gordon at EMD banquet, Detroit, October 18, 1961, AAR.

27. Because of declining locomotive demand, GM transferred the Cleveland plant to the Euclid Division in 1954, although the facility continued to produce diesel locomotive parts. *Railway Age* 124:26 (June 26, 1948), 1306; 129:8 (August 19, 1950), 43–44; 130:25 (June 25, 1951), 96; 137:15 (October 11, 1954), 11–12; EMD *Streamliner* 12:2 (June 18, 1948), GMI, folder 83-12.101; GM-EMD, 1951 Progress Report, John W. Barriger III Collection at the St. Louis Mercantile Association Library (hereafter referred to as the Barriger Collection), box H-34, 15–16.

28. *Railway Age* 122:22 (May 31, 1947), 1110–11.

29. Ibid., 119:4 (July 28, 1945), 179; 122:6 (February 8, 1947), 324–25; 122:22 (May 31, 1947), 1110–11; 122:25A (June 23, 1947), 1294D19–26; 173:4 (August 28, 1972), 40–41; GM-EMD, *The Diesel Locomotive: Revolution on Rails*, ca. 1950, AAR.

30. *Industrial Marketing* 31:11 (November 1946), 30–31; *Railway Age* 173:4 (August 28, 1972), 40–41; John E. Tilford to M. H. Gardner, July 15, 1948, L & N Collection, box 2, folder 81930.

31. *Railway Age* 173:4 (August 28, 1972), 40–41 (quotes); GM-EMD, "Safeguarding Railroad Earnings," 1952, AAR; Nelson C. Dezendorf, speech before the New York Society of Security Analysts, New York, June 1, 1951, AAR; R. C. Parsons to Ivan E. Rice, October 13, 1953, L & N Collection, box 2, folder B-77203.

32. *Railway Age* 125:26 (December 25, 1948), 1188; 134:9 (March 2, 1953), 12; United States of America vs. General Motors Corporation, April 12, 1961, 4.

33. EMD also conducted extensive research into methods of spare-parts packaging. In 1959, EMD installed an IBM punch card Daily Multiple Order Invoice System to coordinate parts distribution at La Grange and eight branch locations. *Fortune* 38 (July 1948), 76–81; *Railway Age* 126:10 (March 5, 1949), 478–79; 127:6 (August 6, 1949), 246–49; 146:9 (March 2, 1959), 23, 26; EMD press release, February 24, 1949, AAR.

34. EMD's competitors were unable to match this investment in rebuilding facilities. Baldwin had only its main plant at Eddystone, along with the facilities operated by Westinghouse. Similarly, ALCo expanded its diesel locomotive rebuilding capabilities by using GE service centers, although ALCo did begin to establish its own parts distribution centers in 1952. Fairbanks-Morse relied on its main plant in Beloit.

Business Week, September 2, 1950, 40–42; *Railway Age* 121:14 (October 5, 1946), 573; 132:5 (February 4, 1952), 20–21; 133:5 (August 4, 1952), 96; 138:18 (May 2, 1955), 13; "Welcome to the Electro-Motive Division of General Motors," ca. 1955, GMI, folder B 3/17.

35. An early advocate of gasoline rail cars, Brookmeyer had served as superintendent of transportation for the Big Four Railroad, a subsidiary of the New York Central. He joined EMC as a sales manager in 1925 and later became general sales manager at EMD. *Railway Age* 132:15 (April 14, 1952), 57–58; 135:25 (December 21, 1953), 14; Reck, *The Dilworth Story*, 99.

36. *Railway Age* 129:7 (August 12, 1950), 76; 129:8 (August 19, 1950), 43–44; 132:7 (February 18, 1952), 13; 132:15 (April 14, 1952), 57–58; 147:11 (September 14, 1959), 67.

37. Other problems plagued ALCo's locomotive production and distribution. Since La Grange was located in Chicago, at the center of the North American railroad network, delivery times tended to be shorter and delivery costs lower than from ALCo's Schenectady plant. Taxes were higher in New York than in Illinois, putting ALCo at a further disadvantage. In addition, ALCo suffered from declining labor productivity, probably caused by outdated manufacturing facilities, and relatively higher wages and fringe benefits. "Telecast Script of James J. Reynolds, Vice-President of Manufacturing, Presenting the Facts Behind the Strike Affecting Alco's Schenectady, Auburn, and Dunkirk Plants," April 1, 1957, ALCo Collection, box 6; Garmany, *Southern Pacific Dieselization*, 221; *New York Times*, October 29, 1943, 29; January 14, 1950, 19; *Railway Age* 119:14 (October 6, 1945), 579; 120:2 (January 12, 1946), 146; 125:15 (October 9, 1948), 693–94; 128:3 (January 21, 1950), 181; *Barron's*, May 30, 1955, 19–20.

38. *Railway Age* 120:2 (January 12, 1946), 146; 122:2 (January 11, 1947), 150.

39. Ibid., 119:9 (September 1, 1945), 388.

40. Ibid., 119:24 (December 15, 1945), 970–72. This "expanded role" for steam locomotives did not materialize, as ALCo produced its last such locomotive less than three years later.

41. Ibid., 119:14 (October 6, 1945), 48a–b.

42. Ibid., 122:1 (January 4, 1947), 110–13.

43. *Railway Mechanical Engineer* 121 (October 1947), 541–42.

44. *Railway Age* 124:25 (June 19, 1948), 1225.

45. ALCo, report of the annual meeting, May 19, 1953, ALCo Collection, box 6; ALCo, press release, December 27, 1962, ALCo Collection, box 13; *New York Times*, March 13, 1942, 27; July 14, 1944, 21; *Railway Age* 133:23 (December 8, 1952), 85.

46. EMD, on the other hand, suffered only one serious strike during the postwar period, the 1945–46 United Auto Workers strike against all GM facilities. GM executives, unlike those at ALCo, were able to reach an accommodation with organized labor that traded steadily rising wages and fringe benefits for workplace stability. As a result, EMD's relative labor harmony, along with its lower wage rates, gave it a strong competitive advantage against ALCo. *New York Times*, December 24, 1949, 23; February 21, 1952, 15; August 28, 1952, 32; April 10, 1954, 34; December 7, 1956, 39; October 24, 1958, 68; July 23, 1960, 41; *Railway Age* 119:22 (December 1, 1945), 906; 120:15 (April 13, 1946), 776; United States Wage Stabilization Board, "In the Matter of the Panel Hearing Conducted Under the Dispute Resolution Pro-

cedures of the Wage Stabilization Board Relating to Terms of a Renewed Contract Between American Locomotive Company and the United Steelworkers of America, CIO, May 21-June 27, 1952" (Washington: U.S. Wage Stabilization Board, 1952), 417, 721–22, 725.

47. *Barron's*, April 9, 1956, 29–30; *New York Times*, June 5, 1943, 11; August 2, 1945, 36; October 5, 1945, 14; October 8, 1945, 30; April 16, 1947, 2; February 14, 1949, 7; April 16, 1949, 8; May 26, 1949, 18; August 19, 1950, 28; March 7, 1951, 33; March 10, 1951, 18; October 23, 1952, 15; November 4, 1952, 20; February 21, 1953, 6; March 9, 1953, 18; March 1, 1955, 18; March 18, 1955, 23; January 17, 1957, 36; March 2, 1957, 42; May 18, 1957, 23; June 10, 1958, 25; January 8, 1959, 11; April 11, 1960, 15; April 25, 1960; *Railway Age* 130:7 (February 12, 1951), 125; ALCo, 1946 and 1952 annual reports, ALCo, "Report of the Annual Meeting," April 17, 1951, ALCo Collection, box 6; United States Wage Stabilization Board, "In the Matter of the Panel Hearing," 399–400, 647.

48. At this time, ALCo's corporate structure consisted of four divisions—Locomotive, Railway Steel Spring, Diesel Engine, and Alco Products—and two wholly owned subsidiaries, the Montreal Locomotive Works and the American Locomotive Export Company. *The Commercial and Financial Chronicle*, October 29, 1945, 2010; June 10, 1946, 3126; *Railway Age* 122:5 (February 1, 1947), 290; *Diesel Power and Diesel Transportation*, December 1947, 66–69.

49. *Railway Age* 121:7 (August 17, 1946), 308; 121:16 (October 19, 1946), 636–41; 123:11 (September 13, 1947), 450–53; 124:24 (June 12, 1948), 1156–58. For additional information on the ALCo PA, see Norman E. Anderson and Chris G. MacDermot, *PA4 Locomotive* (Burlingame, California: Chatham Publishing, 1978).

50. *Railway Age* 120:16 (April 20, 1946), 839.

51. Ibid., 122:25A (June 23, 1947), 1294D21; *Diesel Power and Diesel Transportation*, December 1947, 66–69. In April 1949, ALCo transferred its business in steam locomotive repair parts to the Lima-Hamilton Corporation—an admission that ALCo's role in the steam locomotive field had ended. (*Railway Age* 126:16 [April 16, 1949], 802.)

52. ALCo, 1946 annual report, 5; 1947 annual report, 3; *New York Times*, January 2, 1947, 41; *Barron's*, May 26, 1947, 24.

53. ALCo, 1946 annual report. This statement ignores the opening of EMD's La Grange facility some ten years earlier.

54. "General Layout of the Schenectady Works: November 1, 1946," map file, ALCo Collection, box 13.

55. Summary of remarks by Ralph Budd at the EMD Silver Anniversary Dinner, Chicago, October 24, 1947, AAR; United States of America vs. General Motors Corporation, April 12, 1961, 6; *Our GM Scrapbook*, 124–30.

56. "Statement of William F. Lewis," Senate Hearings, November 9, 1955, 2392.

57. *Diesel Power and Diesel Transportation*, December 1947, 66–69, 92; July 1948, 38–41; ALCo, 1948 annual report.

58. J. M. Budd to Robert S. Macfarlane, May 10, 1955, NP Collection, box 898, file 2981, 137.G.2.8(F).

59. Baldwin first offered turbocharged diesel locomotives in 1946, while EMD did not offer this option until 1948 or 1949. *Railway Age* 121:16 (October 19, 1946), 636–41; *Diesel Railway Traction* 15:348 (May 1961), 191–97; GM-EMD, "Confer-

ence Leader's Outline," Subject III, Unit IB, GMI; Garmany, *Southern Pacific Dieselization*, 161.

60. *Industrial Marketing* 31:11 (November 1946), 30–31.

61. Robert B. McColl, Address to the Railroad Executives' Conference, Schenectady, New York, March 1948, ALCo Collection, box 6; ALCo, 1946 annual report, 3, 11; *Railway Age* 121:16 (October 19, 1946), 636–41; 123:11 (September 13, 1947), 450–52; 124:24 (June 12, 1948), 1156–58; 128:18 (May 6, 1950), 894; *New York Times*, May 5, 1950, 30.

62. ALCo, 1954 annual report, 5; *Barron's*, May 30, 1955, 19–20; *New York Times*, August 11, 1953, 30.

63. Henceforth, "Alco" will refer specifically to Alco Products from 1955 on, while "ALCo" will refer to the broad span of the company's history (pre- and post-1955).

64. ALCo, 1954 annual report, 5, 12–13; Perry T. Egbert, "Letter to all Shareholders," April 20, 1955; ALCo Collection, box 6; Schenectady *Union-Star*, May 6, 1954; *Alco Products Review* 4:4 (Fall 1955), 18–21; 5:1 (Winter 1956), 1; 6:2 (Spring 1957), 24–25; *Barron's*, May 30, 1955, 19–20; *Railway Age* 136:18 (May 3, 1954), 10; 136:24 (June 14, 1954, 16); *New York Times*, May 8, 1954, 27; June 4, 1954, 37; April 20, 1955, 53.

65. *Railway Age* 138:18 (May 2, 1955), 14–15.

66. *Alco Products Review* 4:1 (Winter 1955), 11–16; 4:2 (Spring 1955); 7:1 (Winter 1958), 23; *Railway Age* 138:18 (May 2, 1955), 14–15.

67. *Alco Products Review* 5:2–2 (Spring/Summer 1956), 15; *Railway Age* 140:4 (January 23, 1956), 43–44; 140:9 (February 27, 1956), 24–26.

68. ALCo, 1957 annual report, 3–4; ALCo press release, August 6, 1957, ALCo Collection, box 12; *Railway Age* 142:15 (April 15, 1957), 12 (quote); 143:8 (August 19, 1957), 10. In 1957, as ALCo was divesting itself of a third of its Schenectady plant, EMD was planning a 42 percent expansion of its La Grange facility. *Railway Age* 142:2 (January 14, 1957), 134.

69. F. W. Taylor to G. L. Ernstrom, January 19, 1948; C. H. Burgess to Robert S. Macfarlane, December 6, 1954; both in NP Collection, box 898, file 2981, 137.G.2.8(F); Burgess to Macfarlane, March 14, 1957, NP Collection, box 508, file 1261–9, 137.E.3.8(F).

70. J. M. Budd to Macfarlane, April 20, 1955, NP Collection, box 898, file 2981, 137.G.2.8(F).

71. Ibid.

72. Macfarlane to J. H. Poore, February 28, 1955, NP Collection, box 898, file 2981, 137.G.2.8(F).

73. Budd to Macfarlane, May 10, 1955, NP Collection, box 898, file 2981, 137.G.2.8(F).

74. *Railway Age* 133:1 (July 7, 1952), 148–50; American Institute of Mechanical Engineers, "Diesel-Electric Locomotives in Canada."

75. In 1950 Fairbanks-Morse signed a joint diesel locomotive production agreement with the Canadian Locomotive Company, based in Kingston, Ontario. CanLoCo completed its first diesel in 1951 and maintained sporadic production until 1963, long after Fairbanks-Morse had left the American diesel locomotive industry. Altogether, CanLoCo manufactured fewer than one hundred diesel locomotives to Fairbanks-Morse designs. Neither Baldwin nor Lima manufactured diesel locomo-

tives in Canada. *Barron's* 31 (April 16, 1951), 27; 35 (October 10, 1955), 27–28; *Railway Age* 128:20 (May 20, 1950), 1019; 128:24 (June 17, 1950), 1208; 131:7 (August 13, 1951), 37; 133:1 (July 7, 1952), 148–50; 133:3 (July 21, 1952), 13; 140:4 (January 23, 1956), 15; Garmany, *Southern Pacific Dieselization*, 303; Canadian Locomotive Company, Ltd., "To the Shareholders," May 11, 1950, AAR; Fairbanks, Morse & Company, *Pioneers in Industry*, 66; Charles H. Wendel, *Power in the Past*, Vol. 2, 72–73.

76. *Business Week*, August 19, 1950, 98; *Railway Age* 127:6 (August 6, 1949), 267–68; 129:8 (August 19, 1950), 51–52; 132:6 (February 18, 1952), 54; 133:1 (July 7, 1952), 148–50; 133:11 (September 15, 1952), 15.

77. ALCo purchased the Locomotive and Machine Company of Montreal in 1904 and changed its name to the Montreal Locomotive Works in 1908. By the end of 1973, diesels in service on North American railroads included 1,933 built by GM Diesel in Canada, along with 23,307 built by EMD in the United States, 1,288 built by MLW, and a mere 48 built by CanLoCo. During 1973 GM Diesel sold 135 locomotives, compared to 1,011 for EMD and 49 for MLW. *Business Week*, August 19, 1950, 98; *Railway Locomotives and Cars*, April/May 1974, 16; Kirkland, *The Diesel Builders*, Vol. 2, 18.

78. By 1952, 25 percent of MLW's production was in the form of diesel locomotives, with the remainder being heavy equipment for the petroleum, chemical, and related industries. *Commercial and Financial Chronicle* 169 (January 20, 1949), 312; *Railway Age* 133:1 (July 7, 1952), 148–50; MLW-CGE press release, May 10, 1948, AAR; MLW press release, December 21, 1949, AAR; ALCo 1946 and 1952 annual reports.

79. *Wall Street Journal*, January 22, 1975, 12; January 27, 1975, 4; *Business Week*, March 10, 1975, 81. For additional information on Bombardier and on the importance of government orders to the mass-transit market, see Chris DeBresson and J. Lampel, "Bombardier's Mass Production of the Snowmobile: The Canadian Exception?" *Scientia Canadensis* 29 (1985), 133–49; and Robert Dalpe, Chris DeBresson, and Hu Xiaoping, "The Public Sector as First User of Innovations," *Research Policy* 21 (1992), 251–63.

80. Steven Klepper and Kenneth L. Simons, in "Innovation and Industry Shakeouts," *Business and Economic History* 25 (Fall 1996), 81–89, examine three models relating technological change to industry shakeouts. They label these the innovative gamble theory, the dominant design theory, and the evolutionary theory. In their study of the automobile, tire, television, and penicillin industries, the authors conclude that the evolutionary model best explains observed industry competitive patterns. Even though the locomotive industry contained far fewer producers than any of the industries studied by Klepper and Simons, the same conclusions hold true. A dominant diesel locomotive design had been established well before the shakeout began, and both product and process innovations had likewise slowed prior to the shakeout. Instead, Electro-Motive, as an early entrant and as the dominant producer, could effectively employ R and D programs to steadily distance itself from its smaller competitors. In other words, Baldwin, Lima, Fairbanks-Morse, and, to a lesser extent, ALCo, simply could not keep pace with Electro-Motive's incremental improvements and instead fell further and further behind as the diesel locomotive industry evolved. Market shakeout in the locomotive industry also supports the conclusions of Tushman and Anderson ("Technological Discontinuities and Organizational Envi-

ronments," 460) that "Competence-enhancing discontinuities result in greater product-class consolidation, reflected in relatively smaller entry-to-exit ratios and decreased interfirm sales variability." In other words, as EMD made incremental changes to the dominant technology, railroads tended to standardize on its products and this, in turn, increased firm exits and decreased the variety of locomotive models available.

81. Virtually all of these steam locomotives were destined for export.
82. *Railway Age* 120:15 (April 13, 1946), 765–69.
83. Ibid., 122:26 (June 28, 1947), 1309–10.
84. Ibid., 124:14 (April 3, 1948), 687–88.
85. "Statement of O. DeGray Vanderbilt III," Senate Hearings, 2375.
86. Ibid., 2366.
87. Baldwin Locomotive Works, 1947 annual report.
88. *Railway Age* 124:14 (April 3, 1948), 687–88.
89. L. W. Hance memorandum, March 22, 1948, Baldwin Collection, box C-12–5, carton 6. Albert Hoefer added the written comment.
90. B. G. Mellin to E. J. Harley, August 30, 1946, in "Planning and Simplification Committee," Baldwin Collection, box C-12–5, carton 6, emphasis in the original.
91. F. B. Bowers memorandum, March 1, 1948; "Diesel Engine Cost Reduction Committee-Minutes," both in Baldwin Collection, box C-12–5, carton 6.
92. R. B. Crean memorandum, November 6, 1946, Baldwin Collection, box C-12–5, carton 6.
93. *Railway Age* 117:22 (November 25, 1944), 828; 126:19 (May 7, 1949), 929; *Railway Mechanical and Electrical Engineer*, April 1952, 59–64; J. R. Watt to J. B. Hill, May 28, 1941, L & N Collection, box 1, folder A-15113; G. W. Honsberger to C. J. Bodemer, July 2, 1942, L & N Collection, box 94, folder 51170, part 1B; Baldwin Locomotive Works, press release, June 13, 1949, AAR.
94. Blickle, "The History, Development, Operation, and Performance of Diesel Locomotives on the Santa Fe Railway," ATSF Collection.
95. A. P. Wendland to J. T. Morrison, November 29, 1948, Baldwin Collection, box 1, file BF.34/BF44.
96. *Railway Age* 125:6 (August 7, 1948), 303; Baldwin Locomotive Works, 1948 annual report; Baldwin press release, July 29, 1948, AAR.
97. *Railway Age* 125:6 (August 7, 1948), 303.
98. A. C. Ricketts to D. R. Staples, September 14, 1948; H. S. Swoyer to D. R. Staples, November 7, 1950, both in Baldwin Collection, box C-11–3, carton 3; "Baldwin-Lima-Hamilton Diesel-Electric Locomotives, Bulletin #2002," February 1, 1954, Baldwin Collection, box 47.
99. *The Magazine of Wall Street*, June 30, 1951, 362–63.
100. *Railway Age* 134:12 (March 23, 1953), 122; Baldwin Locomotive Works, 1944 and 1949 annual reports; Baldwin-Lima-Hamilton, 1952, 1953, and 1954 annual reports.
101. *Railway Age* 136:23 (June 7, 1954), 16; Baldwin-Lima-Hamilton, 1954 and 1955 annual reports.
102. Between 1952 and 1954, Baldwin-Lima-Hamilton acquired the Austin-Western Corporation, the Hydropress Company, the Loewy Construction Company, the O. S. Peters Company, the Ruge-deForest Company, and the Madsen Iron

Works. *Business Week*, August 12, 1950, 80; Baldwin-Lima-Hamilton, 1950 and 1954 annual reports.

103. Baldwin-Lima-Hamilton, 1955 annual report. BLH executives may well have read Peter F. Drucker's masterful account of the GM model, *Concept of the Corporation* (New York: The John Day Company, 1946).

104. *The Commercial and Financial Chronicle* 185 (May 16, 1957), 2274 (quote); *Railway Age* 140:2 (January 9, 1956), 204.

105. *The Commercial and Financial Chronicle* 185 (May 16, 1957), 2274; *Railway Age* 145:2 (July 14, 1958), 17; Baldwin-Lima-Hamilton, 1956 annual report.

106. Brown, *The Baldwin Locomotive Works*, 223–28.

107. Ira P. Carnes, "To the President, Officers, and Employees of the Lima Locomotive & Machine Co.," 1912, ACHS; "BLH Observes 100th Year," *Lima News*, February 23, 1969; Richard J. Cook, "Lima: 100 Years of Steam and Steel," *Railroad Model Craftsman*, December 1969, 26–33. In 1903, Lima completed a new facility to fill an increasing number of orders for Shay locomotives. The plant was expanded in 1910, in 1912, and again between 1918 and 1920. In 1912, the company was reorganized as the Lima Locomotive Company and was again reorganized, in 1916, as the Lima Locomotive Works, Incorporated. (O. B. Schultz, "The History of the Lima Locomotive Works," a paper presented at a meeting of the Chicago Chapter of the Railway and Locomotive Historical Society, March 14, 1947, ACHS, box 889, 6–10). For more information on Shay locomotives, see Michael Kock, *The Shay Locomotive: Titan of the Timber* (Denver: World Publishing, 1971).

108. Eric Hirsimaki, *Lima: The History* (Edmonds, Washington: Hundman Publishing, 1986), 113, 136; Earle Davis, "60 Years of Lima Locomotives," *Railroad Magazine*, December 1939, 21; *Railway Age* 79:11 (September 12, 1925), 467–71; 81:3 (July 17, 1926), 101–2; C. B. Peck to A. Lane Cricher, July 9, 1925, Records of the Bureau of Foreign and Domestic Commerce, National Archives (Record Group 151, hereafter referred to as the BFDC Records). Several Lima "Superpower" locomotives still exist, either preserved as static displays or used in excursion service. One such locomotive, built for the Southern Pacific in 1936, was used for the Bicentennial Freedom Train forty years later. (Richard K. Wright, *America's Bicentennial Queen—Engine 4449* (Oakhurst, California: Wright Enterprises, 1975), 28–34, ACHS). For more information on Lima's "Superpower" locomotives, see: Richard J. Cook, *Super-Power Steam Locomotives* (San Marino, California: Golden West Books, 1966); and Eugene L. Huddleston and Thomas W. Dixon, *The Allegheny: Lima's Finest* (Edmonds, Washington: Hundman Publishing, 1982).

109. H. W. Snyder, "The Necessity for Improvement in the Design and Operation of Present-Day Locomotives," paper presented at the spring meeting of the American Society of Mechanical Engineers, May 23–26, 1921, ACHS, box 1063; Robert B. Krueger, Patent for "Locomotive Frame and Engine Supporting Structures," #1786854, issued December 30, 1930; Herbert W. Snyder and Robert B. Krueger, Patent for "Locomotive Steam Distribution," #1826290, issued October 6, 1931; Lima Locomotive Works, Inc., "Engine Specification—Order No. 1070, Shop No. 6883," no date, ACHS, box 1063; Correspondence in ACHS, box 897. David Weitzman provides a fascinating account of steam locomotive construction at Lima in *Superpower: The Making of a Steam Locomotive*.

110. Schultz, "History"; Hirsimaki, *Lima: The History*, 74, 184–87; John E. Dixon

to W. Rodney Long, February 26, 1930, and January 22, 1931, BFDC Records. Hirsimaki (pp. 194–95) reports that Lima had several diesel-mechanical or diesel-electric locomotives under design between 1938 and 1940. I have found no confirmation of this in any other source, however.

111. During World War II, the War Production Board mandated that Lima produce only steam locomotives although, at the time, the company had no intention of manufacturing diesels. O. B. Schultz, "The History of the Lima Locomotive Works," 1947, ACHS, box 889, 6–10; Hirsimaki, *Lima: The History*, 74, 184–87; "Title 32—National Defense, Chapter IX—War Production Board, Subchapter B—Division of Industry Operations, Part 1188—Railroad Equipment, General Limitation Order L-97, April 4, 1942, WPB Records.

112. *Railway Age* 123:2 (July 12, 1947), 51–52; *Railway Mechanical Engineer* 122 (January 1948), 13–15; Jane Bunyan Boehme, "Builders of the Iron Horse," *Columbus Sunday Dispatch Magazine*, July 20, 1947, 6 (quote).

113. Lima-Hamilton, 1948 annual report, 4–5.

114. Lima Locomotive Works, Inc., "Notice of Special Meeting of Stockholders to Act on Plan to Acquire Assets of General Machinery Corporation," August 15, 1947, AAR; Lima-Hamilton Corporation, 1949 annual report, 15; *Railway Age* 107:7 (August 12, 1939), 265; 107:17 (October 21, 1939), 625–27; 126:2 (January 8, 1949), 250–51.

115. Lima, "Notice of Special Meeting of Stockholders."

116. "Interview with Henry A. Rentschler," *The Business History Bulletin* 4 (Summer/Fall 1990), 5–8. Henry Rentschler, the grandnephew of George Rentschler, was in his early twenties at the time of the Lima-Hamilton merger. He recalls that "it was obvious that we [executives from General Machinery] were running that show very quickly." In August 1949, Lima-Hamilton president John E. Dixon resigned his position prior to becoming chairman of the board. Samuel G. Allen resigned from that position. Daniel Ellis, the former vice president in charge of manufacturing at Lima and at Lima-Hamilton, became the new president of the corporation. Lima-Hamilton, 1949 annual report, 3, 6–7; *Railway Age* 127:7 (August 13, 1949), 307; 127:7 (August 20, 1949), 353. Ironically, financial analysts initially claimed that Lima stockholders had been the principal gainers in the merger. *Barron's*, September 1, 1947, 35.

117. *Railway Age* 125:14 (October 2, 1948), 643–44.

118. Lima-Hamilton, 1948 annual report, 5; 1949 annual report, 4–5; *Railway Age* 126:16 (April 16, 1949), 802.

119. *Railway Age* 125:15 (October 9, 1948), 669–70; 127:19 (November 5, 1949), 790–93; 129:10 (September 2, 1950), 63–66.

120. Baldwin-Lima-Hamilton Corporation, Consolidation of Baldwin Locomotive Works and Lima-Hamilton Corporation, December 5, 1950, AAR; Baldwin-Lima-Hamilton, 1950 annual report, 3–5; *Railway Age* 129:7 (August 12, 1950), 75; 129:21 (November 18, 1950), 76; *Business Week*, August 12, 1950, 80; November 11, 1950, 115.

121. Baldwin-Lima-Hamilton, 1951 annual report; "City's Locomotive Era Ends," *Lima News*, September 9, 1951. Baldwin-Lima-Hamilton supplemented the production of the original Lima Shovel and Crane Division when it purchased the Austin-Western Company, a producer of road-building equipment, in December 1950.

NOTES TO CHAPTER VI

122. *Diesel Power and Diesel Transportation* 30:11 (November 1952), 46–49.

123. Louis A. Marre, "Too Little, Too Late," *Railfan* (Winter 1974), 51–58.

124. For example, the following exchange took place at the 1955 hearings of the Senate Subcommittee on Antitrust and Monopoly (p. 2354):

> Mr. NEVILLE: How much money do you have invested in plant and equipment for the manufacture of the locomotives?
>
> Mr. PETERSON [Fairbanks-Morse vice president]: It did not occur to me that you would ask me that; I am afraid that I do not have anything that would be reliable.

125. *The Commercial and Financial Chronicle* 165 (May 5, 1947), 2412–13.

126. *Railway Age* 121:22 (November 30, 1946), 932.

127. Like the other locomotive producers, Fairbanks-Morse suffered during the postwar strike wave—Beloit shut down for 144 days in 1946. *Business Week*, May 17, 1947, 59–64; *Railway Age* 121:4 (July 27, 1946), 126–29; Kirkland, *The Diesel Builders, Vol. 1*, 38.

128. *Railway Mechanical and Electrical Engineer*, May 1950, 281.

129. *Sales Management* 67:7 (October 1, 1951), 120–24.

130. *Railway Age* 126:24 (June 11, 1949), 1179; 127:15 (October 8, 1949), 658; *Steel* 118:2 (January 14, 1946), 73; Fairbanks, Morse & Company, *Pioneers in Industry*, 18, 109.

131. Garmany, *Southern Pacific Dieselization*, 294–300.

132. *Business Week*, May 17, 1947, 59–64; *Railway Age* 90:19 (May 9, 1931), 935; 121:22 (November 3, 1946), 932; 127:4 (July 23, 1949), 191; 146:26 (June 29, 1959), 63; Fairbanks-Morse, press release, May 8, 1944, AAR; Fairbanks, Morse & Company, *Pioneers in Industry*, 74–76, 156–57.

133. *Barron's* 43 (May 6, 1963), 5, 17–19 (quote); *Business Week*, February 16, 1957, 42.

134. For information on British managerial practices, see Leslie Hannah, *The Rise of the Corporate Economy*, 2d ed. (London: Methuen & Company, 1983). Chandler, in *Scale and Scope*, offers a sharply critical view of the debilitating effect of family management on British business.

135. *Barron's* 35 (October 10, 1955), 27–28; 43 (May 6, 1963), 5, 17–19; *Business Week*, April 14, 1956, 116–18; February 16, 1957, 42; May 18, 1957, 48; December 7, 1957, 82; *Forbes* 81 (May 1, 1958), 19; *Railway Age* 136:17 (April 26, 1954), 38–39.

136. *Barron's* 43 (May 6, 1963), 5, 17–19.

137. *Forbes* 81 (May 1, 1958), 19; 89:12 (June 15, 1962), 15–16; 91 (April 15, 1963), 39; Kirkland, *The Diesel Builders, Vol. 1*, 65.

138. *Railway Age* 137:10 (September 6, 1954), 16; 143:12 (September 16, 1957), 7; 147:11 (September 14, 1959), 68; Kirkland, *The Diesel Builders, Vol. 1*, 53, 65, 67.

139. "Statement of V. H. Peterson," Senate Hearings, November 9, 1955, 2356.

140. *Barron's* 31 (April 16, 1951), 27; United States of America vs. General Motors Corporation, April 12, 1961, 7, 10.

141. Fairbanks-Morse would have done well to follow the example of the Ingalls Shipbuilding Company. Although Ingalls announced plans for a complete diesel locomotive line in 1946, it built only one unit before its management realistically assessed the situation and left the industry. Ingalls, its financial resources still intact,

later became a successful defense contractor. Executives at Ingalls realized that their company's skills in custom-built ship production were not applicable to the standardized production of diesel locomotives. Had Fairbanks-Morse executives acted with similar foresight, the company would have been spared a great deal of financial hardship. *Railway Age* 120:21 (May 25, 1946), 1062–64.

142. Chandler's best known works include *Strategy and Structure* (1962), *The Visible Hand* (1977), and *Scale and Scope* (1990). For analyses of the roles played by custom small-batch producers, see Piore and Sabel, *The Second Industrial Divide* (1984); and Scranton, *Figured Tapestry* (1989), and "Diversity in Diversity" (1991).

143. The standardized, mass-production nature of the diesel locomotive industry may be changing. Within the past decade, railroad mergers have led to a small number of giant railroad systems, which, particularly in periods of low locomotive demand, have persuaded both GE and EMD to make substantial modifications to established locomotive designs. These design changes are far less extensive than those that characterized the steam locomotive industry, however.

Chapter VII
The Era of Oligopoly

1. Following the 1901 ALCo merger, the steam locomotive industry itself constituted a near-duopoly, since ALCo and Baldwin clearly dominated the field. Lima frequently had to survive on orders that its much larger rivals could not accommodate. The presence of two controlling firms encouraged Lima to become an innovator in the steam locomotive industry, offering new and improved products that provided the company with a competitive edge over ALCo and Baldwin.

2. *Railway Age* 152:2 (January 15, 1962), 16, 103; Eastern Railroad Presidents Conference, *Yearbook of Railroad Information, 1963 Edition* (Jersey City, N.J.: Eastern Railroad Presidents Conference, 1963), 16, 20.

3. *Railway Age* 152:2 (January 15, 1962), 16, 103; 154:19 (May 20, 1963), 42–44; 158:23 (June 7, 1965), 18–22; *Business Week*, June 12, 1965, 186; GM-EMD, "645 Series General Motors Diesel Engine," ca. 1967, GMI, folder 83–12.101.

4. *Railway Age* 158:23 (June 7, 1965), 18–22.

5. The government's attack on GM and EMD arose from a new interest in antitrust activity after World War II. In 1950, Congress passed the Cellar-Kefauver Act, which allowed the Justice Department to use market share as evidence of monopolistic tendencies. This provision, along with restrictions on the acquisition of firms in the same or related industries, was one factor in the growth of conglomerates in the postwar period.

6. To be fair to the federal government, during the 1960s economists were only beginning to understand the relationship between economies of scale and bigness. The impact of path-breaking books, such as Alfred D. Chandler Jr., *Strategy and Structure* (1962), *The Visible Hand: The Managerial Revolution in American Business* (1977), and John Kenneth Galbraith, *The New Industrial State* (1967), postdated the EMD antitrust suits.

7. Congress, Senate, Subcommittee on Antitrust and Monopoly, *A Study of the Antitrust Laws: Bigness and the Concentration of Economic Power—A Case Study of General Motors Corporation*, 84th Cong., 2d sess., 1956, 44.

8. *Business Week*, November 26, 1955, 32; October 5, 1957, 50; October 24, 1959, 141–42; April 22, 1961, 34–36; *Traffic World* 97:17 (April 28, 1956), 98.

9. *The Journal of Commerce*, October 23, 1961.

10. Stenographer's Minutes, United States of America vs. General Motors Corporation, May 22, 1961; Cleveland *Plain Dealer*, April 14, 1961; *Railway Age* 150:16 (April 17, 1961), 10.

11. United States of America vs. General Motors Corporation, April 12, 1961, 10–15.

12. *Railway Age* 154:2 (January 21, 1963), 11–12; Chicago *Daily News*, January 14, 1963; *Wall Street Journal*, January 15, 1963, 2; December 24, 1964; December 29, 1964.

13. S. T. Stackpole to D. C. McGuire (copies to C. R. Scharff and C. A. Sullivan), September 2, 1937, PRR Collection, box 597, file 416/10.

14. J. B. Hill to E. A. deFuniak, September 6, 1939, L & N Collection, box 1, folder A-15113.

15. Report of Appraisers, Supreme Court of the State of New York, in the matter of the Applications of Samuel Posen, Clara Berdon, Belle Kuller, Edgar Scott, et al., 1944, AAR.

16. The Honorable Sidney Sugarman, Opinion, United States of America against General Motors Corporation, U.S. District Court, Southern District of New York, May 25, 1961.

17. Stenographer's Minutes, United States of America vs. General Motors Corporation, May 22, 1961; Herbert A. Charlson to Roy H. Johnson, May 31, 1961, National Archives, Northeast Region.

18. Leo F. Tierney to Louis W. Menk, June 21, 1967, NP Collection, box 898, file 2981, 137.G.2.8(F).

19. *Railway Age* 138:12 (March 21, 1955), 7; 140:16 (April 16, 1956), 47–49.

20. Minutes of Pennsylvania Railroad monthly diesel meeting, June 15, 1949, PRR Collection, box 602, folder 416.043/11; *Railway Age* 122:5 (February 1, 1947), 293; 122:7 (February 15, 1947), 364–65; 125:18 (October 30, 1948), 814; 132:22 (June 2, 1952), 69; 133:24 (December 15, 1942), 60–62; GE, "General Electric's Traction System Gives You Maximum Locomotive Performance," 1955, AAR; O. L. Dunn, "Transportation and Automation: General Electric in Greater Erie," ca. 1960, Barriger Collection, box H-13.

21. *Railway Age* 126:19 (May 7, 1949), 932; 143:6 (February 9, 1953), 14; 136:9 (March 1, 1954), 11–12; 144:15 (April 14, 1958), 46; GE, "General Electric Motive Parts Service," 1957, AAR; GE press release, April 10, 1958, AAR; GE Locomotive Products Department, Transportation Systems Business Division, "Electric Locomotives," 1971, Barriger Collection, box H-13. GE claimed that major overhauls, which took between four and six weeks, would double the effective life of a locomotive and cost 20–40 percent of the price of a new one.

22. *Railway Age* 127:20 (November 12, 1949), 855–56; 130:13 (April 2, 1951), 59–62; 131:15 (October 8, 1951), 41–43.

23. A conference of railroad locomotive maintenance officers, for example, concluded that "the development of the Alco 244 engine has been a fertile field for research with respect to bearings and bearing failures." "Report of the Committee on Diesel Mechanical," *Proceedings of the 16th Annual Meeting of the Locomotive Maintenance Officers' Association*, September 27, 1954, vol. 1, 107.

24. James P. Newell to A. J. Greenough, April 19, 1957, PRR Collection, box 333, folder 411.043/8.

25. GE, "Marketing Forecast for Road Locomotives," ca. 1954, Barriger Collection, box H-34.

26. GE, "Market Analysis, Domestic Road Locomotives," February 1955, Barriger Collection, box H-34.

27. GE, "Marketing Research and Product Planning Presentation, Locomotive and Car Equipment Department Marketing Conference, September 22, 1954," Barriger Collection, box H-34 (emphasis in the original).

28. GE, "Market Analysis, Domestic Road Locomotives," February 1955, Barriger Collection, box H-34; O. L. Dunn, "Transportation and Automation: General Electric in Greater Erie," ca. 1960, Barriger Collection, box H-13; John C. Aydelott, "A Fresh Approach to Diesel-Electric Locomotive Design," paper presented at the American Institute of Electrical Engineers, Winter General Meeting, New York, January 29–February 3, 1961, Barriger Collection, box H-33.

29. *Railway Age* 148:19 (May 9, 1960), 32–33.

30. Ibid., 138:10 (March 7, 1955), 11; 141:21 (November 12, 1956), 30–32; 148:18 (May 2, 1960), 9; 149:14 (October 3, 1960), 36; 153:3 (July 16/23, 1962), 36.

31. *Business Week*, December 25, 1965, 20; *Modern Railroads* 19:11 (November 1964), 138–41; *Railway Age* 155:11 (September 9, 1963), 36–37; 159:10 (September 13, 1965), 68–70, 107; 163:9 (September 4, 1967), 29; 164:21 (June 3, 1968), 27; Marx, "Technological Change and the Theory of the Firm," 9.

32. J. S. Swan to F. W. Kirchner, June 10, 1960, L & N Collection, box 107, folder 3098; Robert N. Cotton interview, L & N Oral History Project, May 10, 1980, tapes #835–37, 34 (quotes).

33. *Forbes*, May 10, 1982, 105–6.

34. *Wall Street Journal*, September 3, 1993, A1.

35. *The Journal of Commerce*, July 1, 1992, 2B.

36. *Railway Age* 182:13 (July 13, 1981), 18–19; *New York Times*, April 5, 1989, D1.

37. *Railway Age* 172:3 (February 14, 1972), 34; 177:14 (August 9, 1976), 24–25; *Automotive News* 59:4986 (September 16, 1983), 272.

38. *Railway Age* 181:7 (April 14, 1980), 16–24; *Forbes*, May 10, 1982, 105–6.

39. *Railway Age* 183:2 (February 1983), 60B; 183:21 (November 8, 1982), 23–25; 184:8 (August 1983), 51–53; 184:9 (September 1983), 26; *Forbes*, May 10, 1982, 105–6; *Industry Week*, November 29, 1982, 75–76; *Iron Age* 226:15 (May 20, 1983), 74–77; *Wall Street Journal*, September 3, 1993, A1; McCall, "Dieselisation of American Railroads," 16.

40. Using the analytical framework provided by Rebecca Henderson and Kim Clark, it is apparent that EMD had thrived during half a century of incremental innovation which reinforced the division's core concepts and left unchanged the linkages between core concepts and product components (i.e., the product's "architecture"). While architectural innovation had little effect on EMD during the 1970s and 1980s, modular innovation had a devastating impact. As GE improved the reliability of its locomotive models, EMD had difficulty mastering new component technologies, particularly those regarding microprocessor controls and other sophisticated electrical equipment. Henderson and Clark, "Architectural Innovation: The

NOTES TO CHAPTER VII

Reconfiguration of Existing Product Technologies and the Failure of Established Firms."

41. *Railway Age* 185:10 (October 1984), 49; *The Journal of Commerce*, July 3, 1986, 5A (quote); *Wall Street Journal*, September 3, 1993, A1.

42. *Railway Age* 190:8 (August 1989), 57–64; 192:9 (September 1991), 41–48; 195:6 (June 1994), 33–38; *Wall Street Journal*, December 23, 1993, B4.

43. *The Journal of Commerce*, January 11, 1984, 2A; William L. Withuhn, "New Steam in the 1980s—What Happened?" *Railroad History* 156 (Spring 1987): 7–10.

44. Morrison-Knudsen owned a 65 percent share in M-K Rail. *Railway Age* 193:4 (April 1992), 19; 194:7 (July 1993), 59–61; *Business Marketing* 79:2 (February 1994), 3; *The Journal of Commerce*, March 18, 1991, 6B; *Wall Street Journal*, October 17, 1994, A9E (quote); *Fortune* 113:5 (March 3, 1986), 52.

45. It was in these same shops, nearly fifty years earlier, that Pennsylvania Railroad motive power officials worked frantically, if unsuccessfully, to develop a modern steam locomotive that could outperform the new diesels.

46. *Railway Age* 194:10 (October 1993), 12; *The Journal of Commerce*, April 8, 1991, 2B; *New York Times*, November 9, 1994, D5; *Financial Times of Canada*, May 14, 1994, B3.

47. *Railway Age* 192:8 (August 1991), 21; *Business Marketing* 79:2 (February 1994), 3; *The Journal of Commerce*, June 29, 1992, 3B; *Wall Street Journal*, June 27, 1991, C8.

48. *New York Times*, January 9, 1960, 25; *Forbes*, April 15, 1962, 50.

49. Chandler, *Strategy and Structure*, 13–15.

50. "Manufacturing Capabilities, Alco Products, Inc., Latrobe, Pennsylvania," 1963(?), ALCo Collection, box 11; Address by William G. Miller, President, Alco Products, Inc., presented before the New York Society of Security Analysts, August 29, 1963, ALCo Collection, box 28.

51. William G. Miller, "Remarks at the Annual Meeting," April 17, 1962, ALCo Collection, box 6.

52. Alco Products, Inc., press release, November 2, 1959, ALCo Collection, box 13.

53. Roger C. Witherell to Robert C. Hughes, December 3, 1964, ALCo Collection, box 12; O. G. Dellacanonica, "Statistics—Locomotive Export Markets," November 13, 1961, ALCo Collection, box 11; Alco Products, Inc., Press Release, February(?), 1962, ALCo Collection, box 1; Press Release, April 18, 1961, ALCo Collection, box 12; Press Release, November 12, 1964, ALCo Collection, box 29; ALCo, 1960 annual report, 5; 1963 annual report; *Railway Age* 152:15 (April 16, 1962), 33; 155:11 (September 9, 1963), 35; *International Railway Journal*, July, 1962, 38–39.

54. Remarks by William G. Miller at the annual meeting of stockholders, April 21, 1964, ALCo Collection, box 6 (quote); Dick Mann to Roger Witherell, April 16, 1964, ALCo Collection, box 29; Alco Products, Inc., Press Release, March 26, 1963, ALCo Collection, box 28.

55. O. G. Dellacanonica, "Memorandum No. 3: A Brief Survey of Export Locomotive Sales for the First Eleven Months of 1959," December 7, 1959; Dellacanonica, "Survey of Export Locomotive Sales," September 1, 1960; both in ALCo Collection, box 11. According to Dellacanonica, ALCo's manager of export sales, ALCo's five

models were unable to compete with EMD's thirteen, and he called for reduced dependence on standardized designs, coupled with an increase in the number of licensees and fewer restrictions on existing licensees.

56. Chicago *Daily News*, January 14, 1963; *Wall Street Journal*, January 15, 1963, 2; December 24, 1964; December 29, 1964.

57. Alco Products, Inc., News Release, July 17, 1962, ALCo Collection, box 12; *Business Week*, March 10, 1962, 132–36; *Wall Street Journal*, June 16, 1964; *Railway Age*, 164:3 (January 22, 1968), 49–50.

58. Alco Products, Inc., Press Release, March 11, 1963, ALCo Collection, box 28; Worthington Corporation, Press Release, July 20, 1967, ALCo Collection, box 29; *Barron's*, May 20, 1963, 19; *Railway Age*, 154:4 (February 4, 1963), 16–17; 154:22 (June 10, 1963), 29; 155:9 (August 21, 1963), 50–51; 159:3 (July 19, 1965), 10–12; 163:4 (July 31, 1967), 50.

59. C. H. Burgess to Robert S. Macfarlane, August 8, 1963; F. L. Steinbright to Macfarlane, December 14, 1964; both in NP Collection, box 508, file 1261-9, 137.E.3.8(F).

60. Office of the Chief Mechanical Officer to F. L. Steinbright, December 9, 1965, NP Collection, box 508, file 1261-9, 137.E.3.8(F).

61. William G. Miller to Stuart Hall, June 22, 1965, ALCo Collection, box 29; *New York Times*, September 11, 1963, 58; June 4, 1965, 45; *Railway Age*, 159:14 (October 11, 1965), 24–27; 160:2 (January 17, 1966), 51; 162:2 (January 16, 1967), 44–48; 164:3 (January 22, 1968), 49–50.

62. For additional information on the development of conglomerates in the postwar period, see: Jon Didrichsen, "The Development of Diversified and Conglomerate Firms in the United States, 1920–1970," *Business History Review* 46 (Summer, 1972): 202–19; and Charles Gilbert, ed., *The Making of a Conglomerate* (Hempstead, N.Y.: Hofstra University Press, 1972).

63. In 1840 Henry Rossiter Worthington founded the company that bore his name in order to market several of his inventions, including a direct-acting steam pump and the world's first practical water meter. In the early 1900s, as ALCo was consolidating many of the smaller producers in the locomotive industry, Worthington was diversifying into compressors, generators, condensers, feedwater heaters, and steam turbines. By 1964, Worthington produced diesel and gasoline engines, pneumatic equipment, electric motors, construction equipment, and air conditioning equipment. Typescript of a speech to commemorate the 125th anniversary of Worthington Corporation, February 26, 1965, ALCo Collection, box 11; *New York Times*, July 23, 1964, 33; *Forbes*, August 1, 1964, 23.

64. *Forbes*, August 1, 1964, 23.

65. Some of GE's market share gain came at EMD's expense, although Alco certainly suffered the most. *New York Times*, March 8, 1967, 61; Marx, "Technological Change and the Theory of the Firm," 9.

66. *New York Times*, November 28, 1967, 65; January 7, 1969, 49; *Barron's*, August 17, 1970, 26–27; *Railway Age*, 164:3 (January 22, 1968), 49–50; 164:21 (June 3, 1968), 27; O. M. Kerr, *Illustrated Treasury of the American Locomotive Company* (Alburg, Vt.: Delta Publications, 1980) 18, 23.

Bibliography

Primary Sources

Archives and Libraries

The Bureau of Foreign and Domestic Commerce Records, the National Archives, Washington, D.C. (Record Group 151).

National Archives, Northeast Region, New York, New York.

The Reconstruction Finance Corporation Records, the National Archives, Washington, D.C. (Record Group 234).

The War Production Board Records, the National Archives, Washington, D.C. (Record Group 179).

The American Locomotive Company Collection at the George Arents Research Library, Syracuse University.

The Association of American Railroads Library, Washington, D.C.

The Baldwin Locomotive Works Collection at the Pennsylvania State Archives, Harrisburg.

The Eugene Woodruff Industrial Electrification Papers, Historical Collections and Labor Archives, Pattee Library, The University of Pennsylvania.

The General Motors Institute Alumni Foundation's Collection of Industrial History, Flint, Michigan.

The John W. Barriger III Collection of Railroad History at the Library of the St. Louis Mercantile Association.

The Lima Locomotive Works Collection at the Allen County Historical Society, Lima, Ohio.

The Louisville and Nashville Railroad Collection at the University Archives and Record Center, University of Louisville, Louisville, Kentucky (Record Group 123).

Northern Pacific Railway, President's Subject Files, Minnesota Historical Society, St. Paul.

The Pennsylvania Railroad Collection at the Hagley Museum and Library, Wilmington, Delaware (Record Group 1810).

The Records of the Atchison, Topeka, and Santa Fe Railway at the Kansas State Historical Society, Topeka.

Records of the Railroad Museum of Pennsylvania, Strasburg.

Commercial and Trade Journals

Alco Products Review
Alco Review
American Machinist
Automotive Industries
Automotive News
Baldwin Locomotives
Baldwin Magazine

Barron's
Business Marketing
Business Week
Civil Engineering
Coal Age
Collier's
The Commercial and Financial Chronicle
Diesel Power and Diesel Transportation
Diesel Railway Traction
EMD Streamliner
The Engineer
Engineering
Financial Times of Canada
Forbes
Foreign Commerce Weekly
Fortune
General Electric Review
Industrial Marketing
Industry Week
International Railway Journal
Iron Age
The Journal of Commerce
The Magazine of Wall Street
Modern Railroads
Railfan
Railroad Magazine
Railroad Model Craftsman
Railway Age
Railway Locomotives and Cars
Railway Mechanical and Electrical Engineer
Railway Mechanical Engineer
Railway Progress
Scientific American
Steel
Steel Horizons
System
Traffic World
Trains and Travel

Newspapers

Chicago *Daily News*
Cleveland *Plain Dealer*
Columbus (Ohio) *Sunday Dispatch Magazine*
Dunkirk (New York) *Observer*
Lima (Ohio) *Citizen*
Lima (Ohio) *News*
The Lima Star and Republican-Gazette

PRIMARY SOURCES

New York *Journal-American*
New York Times
Schenectady (New York) *Union-Star*
Wall Street Journal

Annual Reports

Alco Products, Incorporated
The American Locomotive Company
The Baldwin-Lima-Hamilton Corporation
The Baldwin Locomotive Works
The Lima Locomotive Works

Interviews

Cotton, Robert N. May 10, 1980, tapes #835–37, University of Louisville, L & N Oral History Project.

Dezendorf, Nelson C. April 6, 1961, The Kettering Archives, 1965 Oral History Project.

Huddle, C. F. November 10, 1964, The Kettering Archives, 1965 Oral History Project.

Osborn, Cyrus R. June 9, 1964, The Kettering Archives, 1965 Oral History Project.

Rentschler, Henry A. "Interview with Henry A. Rentschler," *The Business History Bulletin* 4 (Summer/Fall 1990): 5–8.

Truxell, Clyde W. March 10, 1961, The Kettering Archives, 1965 Oral History Project.

Dissertations

Agnew, Robert J. "Diesel-Electric Locomotive and Railway Employees." Ph.D. diss., Massachusetts Institute of Technology, 1953.

Bingham, Robert Charles. "The Diesel Locomotive: A Study in Innovation." Ph.D. diss., Northwestern University, 1962.

Hydell, Richard P. "A Study of Technological Diffusion: The Replacement of Steam by Diesel Locomotives in the United States." Ph.D. diss., Massachusetts Institute of Technology, 1977.

Marx, Thomas G. "The Diesel-Electric Locomotive Industry: A Study in Market Failures." Ph.D. diss., University of Pennsylvania, 1973.

Norton, Hugh Stanton. "An Economic Survey of the Diesel Locomotive in the Railroad Industry in the United States." Ph.D. diss., George Washington University, 1956.

Park, Donald Kentfield, II. "An Economic Analysis of Innovation and Activity: American Steam Locomotive Building, 1900–1952." Ph.D. diss., Columbia University, 1973.

Government Documents

Clayton Act. Statutes at Large. Vol. 38 (1914).

U.S. Congress. Senate. *Report to Accompany S. 3031 (For Relief of Lima Locomotive Works, Incorporated)*. 75th Cong., 3rd sess., 1938, S. rp. 1605.

U.S. Congress. Senate. Committee on the Judiciary, Subcommittee on Antitrust and Monopoly. *A Study of the Antitrust Laws*. 84th Cong., 1st sess., 1955.

U.S. Congress. Senate. Committee on the Judiciary, Subcommittee on Antitrust and Monopoly. *A Study of the Antitrust Laws: Bigness and the Concentration of Economic Power—A Case Study of General Motors Corporation.* 84th Cong., 2d sess., 1956.

U.S. Congress. House. Committee on the Judiciary, Subcommittee on Antitrust and Monopoly. *Study of Monopoly Power, Part 4: Mobilization Program.* 82d Cong., 1st sess., 1951.

U.S. District Court, Southern District of New York, United States of America vs. General Motors Corporation, case #61CR356, filed April 12, 1961.

U.S. Wage Stabilization Board, *In the Matter of the Panel Hearing Conducted Under the Dispute Resolution Procedures of the Wage Stabilization Board Relating to Terms of a Renewed Contract Between American Locomotive Company . . . and the United Steelworkers of America, CIO, May 21-June 27, 1952* Washington: U.S. Wage Stabilization Board, 1952.

Secondary Sources

Unpublished Papers

Floyd, Leslie A. "Lima and Manufacturing, 1850–1910." Unpublished seminar paper. The Ohio State University, 1987.

Articles

Abernathy, William, and James Utterback. "Patterns of Industrial Innovation." *Technology Review* 80:7 (1978): 40–47.

Aldag, Robert. "Steam vs. Diesel Locomotives." *Railroad History* 167 (Autumn 1992): 148–57.

Aldrich, Mark. "Combating the Collision Horror: The Interstate Commerce Commission and Automatic Train Control, 1900–1939." *Technology and Culture* 34 (January 1993): 49–77.

Anderson, Philip, and Michael L. Tushman. "Managing Through Cycles of Technological Change." In *Managing Strategic Innovation and Change: A Collection of Readings*, edited by Michael L. Tushman and Philip Anderson, 45–52. New York: Oxford University Press, 1997.

Bezilla, Michael. "Pennsylvania Railroad Motive Power Strategies, 1920–1950." *Railroad History* 164 (Spring 1991): 43–52.

Brown, John K., and Samuel M. Vauclain. "Comments on the System and Shop Practices of the Baldwin Locomotive Works." *Railroad History* 173 (Autumn 1995): 102–41.

Bryant, Lynwood. "Rudolf Diesel and His Rational Engine." *Scientific American* 221 (August 1969): 108–17.

———. "The Development of the Diesel Engine." *Technology and Culture* 17 (July 1976): 432–46.

Childs, William R. "The Transformation of the Railroad Commission of Texas, 1917–1940: Business-Government Relations and the Importance of Personality, Agency Culture, and Regional Differences." *Business History Review* 65 (Summer 1991): 285–344.

Christensen, Clayton M. "The Rigid Disk Drive Industry: A History of Commercial and Technological Turbulence." *Business History Review* 67 (Winter 1993): 531–88.

———, and Richard S. Rosenbloom, "Explaining the Attacker's Advantage: Technological Paradigms, Organizational Dynamics, and the Value Network." *Research Policy* 24 (1995): 233–57.

Clayton, Richard. "Strategic Partnering: Why Alco Failed." *Business Quarterly* 57 (Summer 1992): 83–88.

Dalpe, Robert, Chris DeBresson, and Hu Xiaoping. "The Public Sector as First User of Innovations." *Research Policy* 21 (1992): 251–63.

David, Paul A. "Clio and the Economics of QWERTY." *American Economic Review* 75 (May 1985): 332–37.

———. "Understanding the Economics of QWERTY: The Necessity of History." In *Economic History and the Modern Economist*, edited by William N. Parker, 30–49. Oxford: Basil Blackwell, 1986.

———. "Heroes, Herds and Hysteresis in Technological History." *Industrial and Corporate Change* 1 (1992): 129–80.

DeBresson, Chris, and J. Lampel. "Bombardier's Mass Production of the Snowmobile: The Canadian Exception?" *Scientia Canadensis* 29 (1985): 133–49.

Dellheim, Charles. "Business in Time: The Historian and Corporate Culture." *Public Historian* 8 (Spring 1986): 9–22.

———. "The Creation of a Company Culture: *Cadburys*, 1861–1931." *American Historical Review* 92 (February 1987): 13–44.

Didrichsen, Jon. "The Development of Diversified and Conglomerate Firms in the United States, 1920–1970." *Business History Review* 46 (Summer 1972): 202–19.

Diesel, Eugen. "Rudolf Diesel and His Invention." *Engineering* 167 (March 1949): 257, 281–82.

Galambos, Louis. "Looking for the Boundaries of Technological Determinism: A Brief History of the U.S. Telephone System." In *The Development of Large Technological Systems*, edited by Renate Mayntz and Thomas P. Hughes, 135–53. Boulder, Colorado: The Westview Press, 1988.

———. "Theodore N. Vail and the Role of Invention in the Modern Bell System." *Business History Review* 66 (Spring 1992): 95–126.

———. "The Innovative Organization: Viewed from the Shoulders of Schumpeter, Chandler, Lazonick, et al." *Business and Economic History* 22 (Fall 1993): 79–91.

Hamilton, Harold L. "The Development of the Diesel Locomotive," address to the Pacific Railway Club, Los Angeles, June 9, 1949, reprinted in *Proceedings: The Journal of the Pacific Railway Club* 33 (May 1949): 5–22.

Henderson, Rebecca M., and Kim B. Clark. "Architectural Innovation: The Reconfiguration of Existing Product Technologies and the Failure of Established Firms." *Administrative Science Quarterly* 35 (March 1990): 9–30.

Hounshell, David A. "Elisha Gray and the Telephone: On the Disadvantages of Being an Expert." *Technology and Culture* 16 (April 1975): 133–61.

———. "Hughesian History of Technology and Chandlerian Business History: Parallels, Departures, and Critics." *History and Technology* 12 (1995): 205–24.

Kirby, M. W. "Product Proliferation in the British Locomotive Building Industry, 1850–1914: An Engineer's Paradise?" *Business History* 30 (July 1988): 287–305.

Klein, Maury. "Replacement Technology: The Diesel as a Case Study." *Railroad History* 162 (Spring 1990): 109–20.
Klepper, Steven, and Kenneth L. Simons. "Innovation and Industry Shakeout." *Business and Economic History* 25 (Fall 1996): 81–89.
Leonard-Barton, Dorothy. "Core Capabilities and Core Rigidities: A Paradox in Managing New Product Development," In *Managing Strategic Innovation and Change: A Collection of Readings*, edited by Michael L. Tushman and Philip Anderson, 255–70. New York: Oxford University Press, 1997.
Lipartito, Kenneth. "Innovation, the Firm, and Society." *Business and Economic History* 22 (Fall 1993): 92–104.
———. "Culture and the Practice of Business History." *Business and Economic History* 24 (Winter 1995): 1–41.
Loeb, Alan P. "Birth of the Kettering Doctrine: Fordism, Sloanism, and the Discovery of Tetraethyl Lead." *Business and Economic History* 24:1 (Fall 1995): 72–87.
Lytle, Richard H. "The Introduction of Diesel Power in the United States, 1897–1912." *Business History Review* 42 (Summer 1968): 115–48.
Marx, Thomas G. "Technological Change and the Theory of the Firm: The American Locomotive Industry, 1920–1955." *Business History Review* 50 (Spring 1976): 1–24.
McCall, John B. "Dieselisation of American Railroads: A Case Study." *The Journal of Transport History* 6:2 (September 1985): 1–17.
Nelson, Richard R., and Sidney G. Winter. "In Search of a Useful Theory of Innovation." *Research Policy* 6 (1977): 36–76.
Rowlinson, Michael, and John Hassard. "The Invention of Corporate Culture: A History of the Histories of Cadbury." *Human Relations* 46 (March 1993): 299–326.
Reutter, Mark. "The Great (Motive) Power Struggle: The Pennsylvania Railroad v. General Motors, 1935–1949." *Railroad History* 170 (Spring 1994): 15–33.
———. "The Life of Edward Budd, Part 1: Pulleys, McKeen Cars, and the Origins of the *Zephyr*." *Railroad History* 172 (Spring 1995): 5–34.
———. "The Life of Edward Budd, Part 2: Frustration and Acclaim." *Railroad History* 173 (Autumn 1995): 58–101.
Rosenbloom, Richard S., and Clayton M. Christensen. "Technological Discontinuities, Organizational Capabilities, and Strategic Commitments." *Industrial and Corporate Change* 3 (1994): 655–85.
Schumpeter, Joseph. "The Creative Response in Economic History." *Journal of Economic History* 7 (November 1947): 147–59.
Scranton, Philip. "Diversity in Diversity: Flexible Production and American Industrialization, 1880–1930." *Business History Review* 65 (Spring 1991): 27–90.
Sicilia, David B. "Technological Determinism and the Firm." *Business and Economic History* 22 (Fall 1993): 67–78.
Thompson, Gregory. "Misused Product Costing in the American Railroad Industry: Southern Pacific Passenger Service between the Wars." *Business History Review* 63 (Autumn 1989): 510–54.
Tushman, Michael, and Philip Anderson. "Technological Discontinuities and Organizational Environments." *Administrative Science Quarterly* 31 (September 1986): 439–65.

Usselman, Steven W. "Air Brakes for Freight Trains: Technological Innovation in the American Railroad Industry, 1869–1900." *Business History Review* 58 (Spring 1984): 30–50.

———. "Patents Purloined: Railroads, Inventors, and the Diffusion of Innovation in 19th-Century America." *Technology and Culture* 32 (October 1991): 1047–75.

———. "The Lure of Technology and the Appeal of Order: Railroad Safety Regulation in Nineteenth Century America." *Business and Economic History* 21 (1992): 290–99.

———. "From Novelty to Utility: George Westinghouse and the Business of Invention during the Age of Edison." *Business History Review* 66 (Summer 1992): 251–304.

———. "Determining a Middle Landscape: Competing Narratives in the History of Technology." *Reviews in American History* 23 (1995): 370–77.

Utterback, James M., and Fernando F. Suarez. "Innovation, Competition, and Industry Structure." *Research Policy* 22 (1993): 1–21.

Wachhorst, Wyn. "An American Motif: The Steam Locomotive in the Collective Imagination." *Southwest Review* 72 (Autumn 1987): 440–54.

Wise, George. "A New Role for Professional Scientists in Industry: Industrial Research at General Electric, 1900–1916." *Technology and Culture* 21 (July 1980): 408–29.

Withuhn, William L. "New Steam in the 1980s—What Happened?" *Railroad History* 156 (Spring 1987): 7–10.

Books

Allen, Oliver Field. *The Modern Diesel*. New York: Prentice-Hall, 1947.

The American Locomotive Company. *Locomotive Hand-Book*. New York: The American Locomotive Company, 1917.

———. *Symposium on Diesel Locomotive Engine Maintenance*. New York: The American Locomotive Company, 1953.

Anderson, Norman E., and Chris G. MacDermot. *PA4 Locomotive*. Burlingame, Ca.: Chatham Publishing Company, 1978.

Armstrong, John H. *The Railroad: What It Is, What It Does*, rev. ed. Omaha, Nebr.: The Simmons-Boardman Publishing Corporation, 1982.

The Baldwin Locomotive Works. *The Story of Eddystone*. Philadelphia: The Baldwin Locomotive Works, 1929.

Berge, Stanley, and Donald L. Loftus. *Diesel Motor Trains: An Economic Evaluation*. Chicago: The Northwestern University School of Commerce, 1949.

Bijker, Wiebe E., Thomas P. Hughes, and Trevor Pinch. *The Social Construction of Technological Systems: New Directions in the Sociology and History of Technology*. Cambridge: MIT Press, 1987.

Blum, John Morton. *V Was for Victory: Politics and American Culture During World War II*. New York: Harcourt Brace Jovanovich, 1976.

Brown, John K. *The Baldwin Locomotive Works, 1831–1915*. Baltimore: Johns Hopkins University Press, 1995.

Candee, A. H. *Electric Equipment for Railroad Diesel Motive Power*. East Pittsburgh, Pa.: Westinghouse Electric and Manufacturing Company, 1941.

Chandler, Alfred D., Jr. *Scale and Scope: The Dynamics of Industrial Capitalism.* Cambridge: Harvard University Press, 1990.

———. *Strategy and Structure: Chapters in the History of the American Industrial Enterprise.* Cambridge: MIT Press, 1962.

———. *The Visible Hand: The Managerial Revolution in American Business.* Cambridge: Harvard University Press, 1977.

Condit, Carl W. *The Port of New York: A History of the Rail and Terminal System from the Grand Central Electrification to the Present.* Chicago: University of Chicago Press, 1981.

Constant, Edward. *The Origins of the Turbojet Revolution.* Baltimore: Johns Hopkins University Press, 1980.

Cook, Richard J. *Super-Power Steam Locomotives.* San Marino, Ca.: Golden West Books, 1966.

Cummins, C. Lyle, Jr. *Diesel's Engine, Volume One: From Conception to 1918.* Wilsonville, Oreg.: Carnot Press, 1993.

David, Paul A. *Technical Choice, Innovation, and Economic Growth: Essays on the American and British Experience in the Nineteenth Century.* London: Cambridge University Press, 1975.

Dosi, Giovanni. *Technical Change and Industrial Transformation: The Theory and an Application to the Semiconductor Industry.* London: Macmillan, 1984.

Drucker, Peter F. *Concept of the Corporation.* New York: The John Day Company, 1946.

Eastern Railroad Presidents Conference. *Yearbook of Railroad Information, 1951 Edition.* New York: Eastern Railroad Presidents Conference, 1951.

———. *Yearbook of Railroad Information, 1963 Edition.* Jersey City, N.J.: Eastern Railroad Presidents Conference, 1963.

Fairbanks, Morse, and Company. *Pioneers in Industry: The Story of Fairbanks, Morse & Co., 1830–1945.* Chicago: Fairbanks, Morse, and Company, 1945.

Freeman, Christopher. *The Economics of Industrial Innovation.* 2d ed. London: Francis Pinter, 1982.

Foster, Richard. *Innovation: The Attacker's Advantage.* New York: Summit Books, 1986.

Friedel, Robert, and Paul Israel. *Edison's Electric Light: Biography of an Invention.* New Brunswick: Rutgers University Press, 1986.

Galbraith, John Kenneth. *The New Industrial State.* 4th ed. Boston: Houghton Mifflin, 1985.

Garmany, John Bonds. *Southern Pacific Dieselization.* Edmonds, Wash.: Pacific Fast Mail, 1985.

Geertz, Clifford. *The Interpretation of Cultures.* New York: Basic Books, 1973.

General Electric. *Almanac of General Electric Railroad Progress.* Erie, Pa.: General Electric Transportation Systems Division, 1972.

General Motors Corporation. *The Locomotive Industry and General Motors.* New York: Bar Press, 1973.

———. Electromotive Division. *Diesel War Power: The History of Electro-Motive's Diesel Engines in the Service of the United States Navy.* GM-EMD, 1945.

Gilbert, Charles, ed. *The Making of a Conglomerate.* Hempstead, N.Y.: Hofstra University Press, 1972.

SECONDARY SOURCES

Hannah, Leslie, ed. *Management Strategy and Business Development: An Historical and Comparative Study*. London: Macmillan, 1976.
———. *The Rise of the Corporate Economy* 2d ed. London: Methuen & Company, 1983.
Healy, Kent T. *Performance of the U.S. Railroads Since World War II: A Quarter Century of Private Operation*. New York: Vantage Press, 1985.
Hirsimaki, Eric. *Lima: The History*. Edmonds, Wash.: Hundman Publishing, 1986.
Hounshell, David A. *From the American System to Mass Production, 1800–1932: The Development of Manufacturing Technology in the United States*. Baltimore: Johns Hopkins University Press, 1984.
———, and John Kenly Smith, Jr. *Science and Corporate Strategy: DuPont R&D, 1902–1980*. Cambridge: Cambridge University Press, 1988.
Huddleston, Eugene L., and Thomas W. Dixon. *The Allegheny: Lima's Finest*. Edmonds, Wash.: Hundman Publishing, 1982.
Hughes, Thomas Parke. *Elmer Sperry: Inventor and Engineer*. Baltimore: Johns Hopkins University Press, 1971.
———. *Networks of Power: Electrification in Western Society*. Baltimore: Johns Hopkins University Press, 1983.
Ingham, John N. *Making Iron and Steel: Independent Mills in Pittsburgh, 1820–1920*. Columbus: Ohio State University Press, 1991.
Kalmbach Publishing Company. *Our GM Scrapbook*. Milwaukee: The Kalmbach Publishing Company, 1971.
Kerr, O. M. *Illustrated Treasury of the American Locomotive Company*. Alburg, Vt.: Delta Publications, 1980.
Kirkland, John F. *Dawn of the Diesel Age: The History of the Diesel Locomotive in America*. Glendale, Ca.: Interurban Press, 1983.
———. *The Diesel Builders, Vol. 1: Fairbanks-Morse and Lima-Hamilton*. Glendale, Ca.: Interurban Press, 1985.
———. *The Diesel Builders, Vol. 2: American Locomotive Company and Montreal Locomotive Works*. Glendale, Ca.: Interurban Press, 1989.
Kock, Michael. *The Shay Locomotive: Titan of the Timber*. Denver: World Publishing, 1971.
Latour, Bruno. *Science in Action: How to Follow Scientists and Engineers Through Society*. Cambridge: Harvard University Press, 1987.
Lazonick, William. *Business Organization and the Myth of the Market Economy*. Cambridge: Cambridge University Press, 1991.
Leslie, Stewart W. *Boss Kettering*. New York: Columbia University Press, 1983.
McCurdy, Howard E. *Inside NASA: High Technology and Organizational Change in the U.S. Space Program*. Baltimore: Johns Hopkins University Press, 1993.
National Railway Historical Society, Lehigh Valley Chapter. *History of Mack Rail Motor Cars and Locomotives*. Allentown, Pa.: The National Railway Historical Society, 1959.
Nelson, Richard R., and Sidney Winter. *An Evolutionary Theory of Economic Change*. Cambridge: Harvard University Press, 1982.
Noble, David F. *America by Design: Science, Technology, and the Rise of Corporate Capitalism*. New York: Oxford University Press, 1979.
Olmstead, Robert P. *The Diesel Years*. San Marino, Ca.: Golden West Books, 1975.

Perrett, Geoffrey. *Days of Sadness, Years of Triumph: The American People, 1939–1945.* Madison: University of Wisconsin Press, 1985.

Peters, Thomas, and Robert H. Waterman. *In Search of Excellence.* New York: Harper and Row, 1982.

Pinkepank, Jerry A. *The Second Diesel Spotter's Guide.* Milwaukee: Kalmbach Books, 1973.

Piore, Michael J., and Charles F. Sabel. *The Second Industrial Divide: Possibilities for Prosperity.* New York: Basic Books, 1984.

Proceedings and Committee Reports of the Fifty-Third Annual Meeting of the Association of Railroad Superintendents. Topeka: Hall Lithographing Company, 1949.

Proceedings and Committee Reports of the Fifty-Sixth Annual Meeting of the American Association of Railroad Superintendents. Topeka: Hall Lithographing Company, 1952.

Reck, Franklin M. *The Dilworth Story.* New York: McGraw-Hill, 1954.

———. *On Time: The History of the Electro-Motive Division of General Motors.* Detroit(?): General Motors, Electro-Motive Division, 1948.

Reich, Leonard S. *The Making of American Industrial Research: Science and Business at GE and Bell, 1876–1926.* Cambridge: Cambridge University Press, 1985.

Rosenberg, Nathan. *Inside the Black Box: Technology and Economics.* Cambridge: Cambridge University Press, 1982.

———. *Exploring the Black Box: Technology, Economics, and History.* Cambridge: Cambridge University Press, 1994.

Schumpeter, Joseph. *Capitalism, Socialism, and Democracy.* New York: Harper, 1942.

Scranton, Philip. *Figured Tapestry: Production, Markets, and Power in Philadelphia Textiles, 1885–1941.* Cambridge: Cambridge University Press, 1989.

———. *Proprietary Capitalism: The Textile Manufacture at Philadelphia, 1800–1885.* Cambridge: Cambridge University Press, 1983.

Sloan, Alfred P., Jr. *My Years with General Motors.* New York: Doubleday and Company, 1964.

Smith, Marvin W. *Samuel Vauclain: Courageous Pioneer, Believer in America!* New York: Newcomen Society in North America, 1952.

Thomas, Donald E., Jr. *Diesel: Technology and Society in Industrial Germany.* Tuscaloosa: University of Alabama Press, 1987.

Thompson, Gregory. *The Passenger Train in the Motor Age: California's Rail and Bus Industries, 1910–1941.* Columbus: Ohio State University Press, 1993.

Tiffany, Paul A. *The Decline of American Steel: How Management, Labor, and Government Went Wrong.* New York: Oxford University Press, 1988.

Vauclain, Samuel M. *Mass Production within One Lifetime.* Princeton: Princeton University Press, for the Newcomen Society, 1937.

Vrooman, David M. *Daniel Willard and Progressive Management on the Baltimore & Ohio Railroad.* Columbus: Ohio State University Press, 1991.

Weitzman, David. *Superpower: The Making of a Steam Locomotive.* Boston: David R. Godine, 1987.

Wendel, Charles H. *Power in the Past, Vol. 2: A History of Fairbanks, Morse, and Company.* Atkins, Iowa: Old Iron Book Company, 1982.

SECONDARY SOURCES

Westinghouse Electric and Manufacturing Company. *Forty Years Ago: Being a Brief Account of the History and Accomplishments of the Westinghouse Electric and Manufacturing Company*. East Pittsburgh, Pa.: The Westinghouse Electric and Manufacturing Company, 1924.

―――. *Westinghouse-Equipped Diesel-Electric Locomotives—10 to 80 Tons*. East Pittsburgh, Pa.: Westinghouse, 1942; Reprint, Canton, Ohio: Railhead Publications, 1980.

White, John H., Jr. *The American Railroad Passenger Car*. Baltimore: Johns Hopkins University Press, 1978.

Wright, Richard K. *America's Bicentennial Queen—Engine 4449*. Oakhurst, Ca.: Wright Enterprises, 1975.

Index

Alco Products Incorporated, 140–44; diesel locomotive production at, 108, 142; and exit from the locomotive industry, 144; finances of, 140, 143; and foreign markets, 141–42, 199n.55; and joint production with General Electric, 132–34, 145; manufacturing facilities at, 108; and merger with the Studebaker Corporation, 144; and merger with the Worthington Corporation, 143; personnel at, 140; quality control problems of, 133–34; subsidiaries of, 198. *See also* American Locomotive Company

Allen, Samuel G., 61, 194n.116

American Locomotive Company (ALCo), 59–63, 83–85, 103–9; in Canada, 110–11; corporate reorganization of, 108; diesel locomotive production at, 27, 59–60, 83–84, 105–6; and diversification, 105–6, 108; finances of, 4, 61–62, 107–8; and joint production with General Electric, 25, 27, 60, 86, 107; manufacturing facilities at, 61, 85, 106, 176n.9; marketing efforts at, 85, 104, 107; and McIntosh and Seymour Company, 59–60, 104; origins of, 59; personnel at, 83, 103–5, 145; steam locomotive production at, 60, 83, 107, 176n.5, 189n.51; subsidiaries of, 61, 105–6; during World War II, 75–77, 83–85, 146–47. *See also* Alco Products Incorporated; Worthington Corporation

Atchison, Topeka, and Santa Fe, 16–18, 35, 47, 50, 75, 79, 98, 134, 180n.1

Baldwin-Lima-Hamilton, 116. *See also* Baldwin Locomotive Works and Lima Locomotive Works

Baldwin Locomotive Works, 63–73, 87–89, 112–17; and De La Vergne, 70–71; and diesel engine technology, 70–71; diesel locomotive production at, 65, 68–69, 71–72, 87, 112–14; and exit from the locomotive industry, 116–17; finances of, 4, 65, 69–70, 112, 183n.51; gasoline locomotive production at, 65; and joint production with Westinghouse, 28–29, 31, 64, 68, 87–89, 114–15; manufacturing facilities at, 64–65, 87, 177n.27; marketing efforts at, 114; and merger with Lima-Hamilton, 116; origins of, 63–65; personnel at, 61–64, 66–70, 72–73, 89, 115, 117; steam locomotive production at, 63–64, 66, 87, 112; subsidiaries of, 65–66, 192n.102, 194n.121; technological innovations at, 64–65; during World War II, 75–77, 87–89

Baldwin, Matthias W., 63–64
Barriger, John W., III, 92
Blum, John Morton, 79
Bombardier, Ltd., 111
Brinley, Charles E., 70, 88–89
Brown, John, 12
Budd, Ralph, 43, 53, 56, 77–78, 146, 149–51

Chandler, Alfred D., Jr., 6, 125, 140, 147, 156n.6, 157nn. 13 and 15
Chicago, Burlington, and Quincy Railroad, 35, 43–44; and the Burlington *Zephyr*, 43–46, 151
Christensen, Clayton M., 7
Clayton Act, 11, 20, 129–30
corporate culture, 7–8, 21, 148, 150–52, 159nn. 26 and 27, 159n.28; at ALCo, 62–63, 73–74, 145; at Baldwin, 66–68, 72–74, 117; at Electro-Motive, 36, 102, 149–50, 170n.11

De La Vergne. *See* Baldwin Locomotive Works
Dezendorf, Nelson C., 82
Dickerman, William C., 62–63, 85, 103, 130–31
Diesel, Rudolf, 14–15
Dilworth, Richard, 26, 32–33, 50–51, 82, 102, 168n.47

Egbert, Perry T., 104–5, 144, 148
Electro-Motive Company (EMC), 31–34, 41–56; and design standardization, 33, 49–52; diesel locomotive production at, 35, 43–46, 48, 56; finances of, 55–56, 171n.25; marketing efforts at, 33–36, 51–55, 148–49; manufacturing facilities at, 46, 48–49; personnel at, 31, 47, 52; and purchase by General Motors, 41–42; and railcar production, 23, 31–36; and railcar sales, 35, 43–46; research

Electro-Motive Company *(cont.)*
  and development at, 46–48, 55–56. *See also* General Motors Corporation, Electro-Motive Division
Electro-Motive Engineering Corporation. *See* Electro-Motive Company
engines, diesel, 14–15, 32–33, 25, 29, 38–41, 46–47, 75. *See also individual manufacturers*

Fairbanks-Morse, 89–92, 121–24; in Canada, 190n.75; diesel engine development at, 90–91, 122; diesel locomotive production at, 91–92, 121–22, 124; and exit from the locomotive industry, 124; finances of, 121, 124; and joint production with General Electric, 121–22; management at, 92, 123–24; manufacturing facilities at, 91–92, 121–22; marketing efforts at, 122–23; origins of, 90; and the Penn-Texas Corporation, 124; and renamed the Fairbanks Whitney Corporation, 124; during World War II, 78, 89–92
Fairbanks-Whitney Corporation. *See* Fairbanks-Morse
first-mover advantages, 7, 37, 56, 59, 67, 73, 84, 129, 151, 157n.13
Fraser, Duncan W., 63, 103, 106

Galbraith, John Kenneth, 147
General Electric Company, 24–28, 86, 132–40; and diesel engine technology, 25; diesel locomotive production at, 18, 25, 27–28, 86, 132, 134–35, 137–38; and electrical equipment technology, 24–25, 28, 46–47; finances of, 25, 134; and joint production with American Locomotive, 25, 27, 60, 86, 107, 132–34, 174n.55; and joint production with Fairbanks-Morse, 121–22; manufacturing facilities at, 25, 27, 86, 132–33, 135, 137; marketing efforts at, 107, 133–35; quality control problems at, 135–37; and railcars, 23–26; during World War II, 86, 132
General Machinery Corporation. *See* Lima Locomotive Works
General Motors Corporation; and antitrust litigation, 129–32, 147, 196n.5; and diesel engine research and development, 38–41, 42–43, 171n.20, 173n.50, 181n.16; and proposed sale of EMD, 139–40
General Motors Corporation, Electro-Motive Division (EMD), 80–82, 99–102, 127–32, 136–40; in Canada, 110; and design standardization, 19–20; diesel locomotive production at, 99–100, 110, 128–29, 137–38, 143; and electrical equipment technology, 25; finances of, 76, 96, 139–40; manufacturing facilities at, 20, 80–81, 99–100, 110; marketing efforts at, 20, 22, 100–102; personnel at, 81–82, 99–102, 110, 138; and production standardization, 80–81; quality control problems at, 138; during World War II, 75–82, 147. *See also* Electro-Motive Company
General Motors Diesel, Ltd., 110, 139, 144. *See also* General Motors Corporation, Electro-Motive Division

Hamilton, Harold L., 31–36, 41, 43, 47, 53, 77–78, 82, 146, 148–52, 171n.23
Houston, George H., 66
Hughes, Thomas Parke, 6, 156n.6, 157n.11

Ingersoll-Rand Company, 26–27, 166nn. 14 and 18
innovation; and core capabilities, 6, 61, 157n.12, 159n.27, 198n.40; and government policy, 26, 41, 56, 59, 75–79, 93, 146–47, 151, 166n.12; and incremental technological change, 6, 8, 40, 62–65, 118, 148, 158n.22, 198n.40; and organizational routines, 149–50; and radical technological change, 3–4, 6–7, 10, 36, 125, 146–48, 150, 158n.22, 159n.23; and reverse saliences, 6, 15, 25, 138, 163n.18; and serendipity, 4–5, 56–57, 117, 150, 156n.9; in the steam locomotive industry, 13–14; and technological linkages, 44–45; and value networks, 7, 164n.37

Justice, U.S. Department of, 6, 127, 129–32, 142, 147, 151, 155n.2

Kaufman Act, 26, 28, 68
Kelly, Ralph, 89, 115
Kettering, Charles F., 38–43, 46–47, 123, 146, 149, 151
Kettering, Eugene, 39, 42, 47, 123
Kruttschnitt, Julius, 15

labor, 12, 105
Lima-Hamilton Corporation. *See* Lima Locomotive Works

INDEX

Lima Locomotive Works, 117–21; diesel locomotive production at, 118–20; and merger with the Baldwin Locomotive Works, 120; and merger with the General Machinery Corporation, 119; origins of, 117–18, 193n.107; steam locomotive production at, 118–20; and technological innovation, 117–18, 196n.1; during World War II, 76–77

locomotives, diesel in Canada, 109–11, 190n.75; cost savings associated with, 15–18, 43, 49, 53, 97–98, 163n.22; and standardization, 3, 19–21, 49–52, 71, 80–81, 86, 111, 113, 125–26, 132, 145, 148, 150, 196n.143; financing of, 20, 52–53; and labor, 18–19, 21, 50, 151, 161n.31, 164n.33; manufacturing methods of, 19, 20–21, 48–49; and market shares, 60, 79, 96, 124, 135, 138, 142, 144; marketing of, 21; and number ordered or in service, 37, 75–76, 95, 104, 128, 137, 142–44, 191n.77; operating characteristics of, 15–19, 26. *See also individual manufacturers*

locomotives, steam as adjunct technology, 58–59, 125; financing of, 11–12, 20; manufacturing methods of, 10–12, 19, 161n.4; and market share, 10–11; and marketing efforts, 11–12, 21; and number ordered or in service, 13, 95, 104, 180n.2; operating characteristics of, 13, 16–19. *See also individual manufacturers*

Louisville and Nashville Railroad, 55, 77–78, 85, 88

loyalty, of customers, 21, 34, 58, 89, 99, 112–13, 117, 148, 186n.18

manufacturing, custom (small-batch), 6, 8, 10–11, 12, 19, 61, 68, 111, 117, 125–26, 140, 145, 150, 196n.143

McColl, Robert, 83, 103–5

McIntosh and Seymour Company. *See* American Locomotive Company

McKeen, William R., Jr., 24

monopoly, 6, 99, 109, 127, 129–32, 147, 196nn. 5 and 6

Montreal Locomotive Works, 59, 61, 63, 106, 110, 144, 191n.78. *See also* American Locomotive Company

Morrison-Knudsen Corporation, 139

Nelson, William O., 82
New York Central System, 26–28

organizational synthesis, 151–52
Osborn, Cyrus R., 81–82

patents, 5, 14–15, 25, 45, 48, 117–18, 157nn. 15 and 16

Pennsylvania Railroad, 20, 29, 66, 69, 85, 96–99, 131, 133, 185n.6

Piore, Michael, 125, 150

railcars, 23–36, 165n.1; and the Electro-Motive Company, 31–36; and Fairbanks-Morse, 91; and General Electric, 24–26, 165n.9; production of, 24–25, 30, 35, 91; and Westinghouse, 28–29

Rentschler, George A., 119–20
Rosenbloom, Richard S., 7

Sabel, Charles, 125, 150
Salisbury, Carl, 32, 39, 47, 170n.15
Schumpeter, Joseph, 4
Scranton, Philip, 125, 150
Sloan, Alfred P., Jr., 38–39, 43, 46, 152, 171n.23
Smith, Marvin W., 115, 120
social constructivism, 150–51, 157n.14, 160n.29
Southern Pacific Railroad, 15, 85, 122–23

technological determinism, 5, 150, 157n.14

Vauclain, Samuel, 28, 64, 66–70, 117, 123, 148

War Production Board, 75–79, 80, 83–84, 91, 93, 117, 146–47

Westinghouse Electric and Manufacturing Company, 28–31; and diesel engine technology, 29–30, 70; and diesel locomotive production, 29–31; and electrical equipment technology, 28–29; and joint production with Baldwin, 28–29, 31, 87–89, 114–15; and railcars, 23, 28–29

Whitcomb Locomotive Works, 65, 177nn. 24 and 25. *See also* Baldwin Locomotive Works

Winton Engine Company, 32–33, 39–41, 47–48, 168n.49, 170n.15. *See also* Electro-Motive Company and General Motors Corporation

Worthington Corporation, 143–44, 200n.63. *See also* Alco Products Incorporated